Beautiful Flesh

Beautiful Flesh A BODY OF ESSAYS

Edited by Stephanie G'Schwind

The Center for Literary Publishing
Colorado State University

page x: "Hearts and Bones," © 1982 Words and Music by Paul Simon.

Because this page cannot accommodate all copyright notices, pages
245–247 constitute an extension of the copyright page.

The Center for Literary Publishing
9105 Campus Delivery
Department of English
Colorado State University
Fort Collins, Colorado 80523-9105
coloradoreview.colostate.edu

Typeset in Sabon by the Center for Literary Publishing
Manufacturing by Integrated Books International
Printed in the United States

Library of Congress Cataloging-in-Publication Data

Names: G'Schwind, Stephanie, editor.
Title: Beautiful flesh : a body of essays / edited by Stephanie G'Schwind.
Description: Fort Collins, Colorado : Center for Literary Publishing,
    Colorado State University, [2017] | Includes bibliographical references
    and index. Identifiers: LCCN 2017008187 (print) | LCCN 2017012982
    (ebook) | ISBN 9781885635587 (electronic) | ISBN 9781885635570
    (pbk. : alk. paper) Subjects: LCSH: American essays--21st century. |
    Human body in literature. | LCGFT: Essays.
Classification: LCC PS689 (ebook) | LCC PS689 .B43 2017 (print) | DDC
    814/.6--dc23
LC record available at https://lccn.loc.gov/2017008187

The paper used in this book meets the minimum requirements of the American
National Standard for Information Sciences-Permanence of Paper for Printed
Library Materials, ANSI Z39.48-1984.

1 2 3 4 5   21 20 19 18 17

Contents

STEPHANIE G'SCHWIND

Introduction

IN THE BEGINNING: heart and bone.

Still exhilarated by the thrill of putting together my first nonfiction anthology, *Man in the Moon: Essays on Father and Fatherhood,* I began to look for another project. It didn't take long to find the inspiration: in *Colorado Review,* the literary magazine I edit with my colleagues and several amazing graduate students at Colorado State University, we'd recently published Matthew Ferrence's "Mos Teutonicus," titled for the ancient practice of removing flesh from bones in order to transport a human body long distances, and Katherine E. Standefer's "Shock to the Heart, Or: A Primer on the Practical Applications of Electricity," on her experience of living with an internal cardiac defibrillator. And there it was: an anthology of essays on the body! It would embrace the body's whole, its parts, its form, its function, its spirit, its nature—exploring through many voices a variety of concerns and obsessions—in memoir, personal essay, and lyric essay, inviting a range of traditional and experimental approaches. I put out the call to the creative nonfiction community, and the essays began to come in. Soon I had eyes, and then a nose, and then blood. And here, the focus changed ever so slightly. With an oblique nod to Victor Frankenstein, I decided to build a body. Out of essays.

So collected here are twenty essays, each taking on a different anatomical part, combining to create a body of neither one gender nor one ethnicity. Neither one ability nor one shape. It is a body that celebrates and grieves, remembers and forgets, injures and heals, loves and laments itself. Between a full head of hair rendered patchy by radiation and a pair of feet that run from a troubled marriage, we come to know pain and possibility, compassion and beauty: a virus that wreaks havoc with a man's ability to think, the strange design of sinuses, a woman's tentative love for her teeth, another's urge to bite the flesh from her fingers, the way a father can't resist nibbling his sweet daughter's ear, and another man's reasons for undergoing a vasectomy. Assembled here, there is also blood and breath, skin and spine, knees, ovaries, and belly. And, of course, a pancreas— once, very long ago, called the *kalikreas,* or "beautiful flesh."

Every anthologist assembles a body of some kind. This one is, of course, a human body. But in the gathering and arranging, it wasn't Frankenstein's voice I heard. From the start, beginning with those first two essays, it was Paul Simon's: *You take two bodies* [or twenty] *and you twirl them into one / Their hearts and their bones / And they won't come undone.*

Hearts and bones indeed.

Beautiful Flesh

VICKI WEIQI YANG

Field Notes on Hair

And nothing is less worthy of a thinking man than to see death as a slumber.
Why a slumber, if death doesn't resemble sleep?

—Fernando Pessoa, *The Book of Disquiet*

AFTER THE BRAIN thing[1] the world became divided spatially and temporally. There were those who knew the truth about my illness, and those who knew the easy-to-swallow version that I had personally lubricated for them. And, as much as I tried to prevent this, there was the cleavage of my life's short timeline into two separate but unequal segments: before the brain thing, when I possessed coveted big-name qualities like Radicalism and Bright-Eyed Naïveté; after the brain thing, when I lost a little bit of those things, and also, for several months, a lot of hair.

I suppose I should say now that this isn't a sob story. Within these lines you will find no hysteric account of how the world has wronged me or marred my idyllic youth. If I think it, I refuse to verbalize it and I—I don't think it in those terms except at my lowest of lows. But still, I carry that scar within my head. There are phantom visions for me, the way other people have to live with phantom limbs, and I stand, divided in the wake of a Capgras delusion in which another me had replaced myself without due warning. And the encounter, spanning the length of a midwestern winter, left behind cold streaks of clarity in an otherwise muddled mind.

---

1 Among friends, I almost always refer to it as such. To do otherwise would endow it with undue weight.

Hair loss begins and ends on its own terms.

The whole affair began on a winter weekday, the kind of week-day that, because the school term had come to an end and the absence of the usual regimen left blank vast stretches of time, blended seamlessly from Monday to Friday to Monday. At three a.m. I said goodbye to Tonya, whom I had been talking to online, and then I had my stroke. For some reason I like the sound of that—*I had my stroke,* as though I am talking about some private but routine gesture: in the afternoon, I had my coffee, and at the stroke of three (as if on cue, by clockwork) I had my stroke. Tonya, a quiet, warm-palmed Mexican Jew with perpetual bedhead, was one of my college friends who got the lubricated version. She liked to dress nicely, in a cardigan and loose button-down, for the girls she could picture as girlfriends. If you had a bad cold, she'd make you chicken noodle soup and feed it to you too. That's the kind of girl Tonya was: A girl who tripped over your softest falls. A girl who would apologize incessantly for things that were nobody's fault. That's why I could never tell her the truth about any of it, you understand— not about the nature of my illness, nor about the fact that she almost became the last person to see me alive.

So I had my stroke (a splitting headache, waves of nausea followed by vomiting, residual pangs that made me knead my temples hard). When it happened, I didn't realize that it was a stroke. Strokes weren't *had* by nineteen-year-olds. I consulted the wisdom of the Internet with search terms like *differences between stroke and migraine* and came to the unhelpful conclu-sion that the two often shared symptoms. I tried to go to bed and in the morning had an MRI, which revealed whole areas in my circulatory system that were missing capillaries. So it was back to the blinding white of Chicago.

No, not really white. Light slate.

I moved into a shared room with a high school girl whose digestive organs were failing by the day. I heard her moaning through the heavy curtain between us, until I was transferred to a private room. Occasionally I would lumber about the halls with my own leash in one hand and the other clutching the train of my dress: a blushing bride in green with IV infusion sets as my bridesmaids. The halls were filled with construction paper posters for the sake of false cheeriness. NEUROSURGERY WARD WINS AWARD FOR BEST CARE, one might proclaim. Tests, some involving blood and others involving spinal fluid but all of them involving needles, became a routine part of my day (today, I *had* a CT scan; yesterday, I *had* a lumbar puncture). The nurses clapped hands and complimented me for my "skinny" frame (for an absence of flesh folds made finding veins relatively simple; even so, we were running out of them by the second week). I began looking forward to morphine injections, which made the senses melt deliciously like diluted watercolors.

All of this I learned to take in with serene indifference. I cried at first, because none of the doctors could give a definitive answer. And the radiologist (a studious-looking man with kindly but overworked gray strands) kept averting his eyes. Of all the possible long-term side effects, I dreaded this the most: "a decrease in intellect." What kind of woman would I be without my intellect? Not a beautiful one, I knew. At night I tossed and turned as much as I could without disturbing the plastic needle planted in the crook of my arm. My hair pressed against my face and made it greasy. In the daytime I tried to keep my mind occupied by reading about the upbringing of John Stuart Mill. (He was, of course, infinitely more precocious than I was. Things like this used to make me agonize.)

Embolization One. I took uneasy trips into the REM world and brought back souvenirs. I dreamt of a North Korean footrace through mountains and classrooms with shattered windows. People stepped in the beads of glass dotting the linoleum and hardly noticed the red footprints they left behind. There were many casualties. The winner was awarded a large medal of yellow brass. If you looked closer, it probably came with a raised etching of the Great Leader—I couldn't tell, since one does not zoom in on dreams.

I came in second, or maybe sixth. I remember there was a small lecture to the ten of us forerunners. The speaker was throwing

out jagged pieces of Confucian maxims at us, except he attributed them to Kim Il-sung and I'm not sure if he knew the difference. There was a bony lady (the nurses would have no trouble
finding her veins) who had an especially pious countenance, who
kept bowing little solemn bows and repeating, "That's very wise,
Professor. Ah, yes. That's very wise, Professor."

In the waking world a mere block away, my friends were cel-
ebrating the cancellation of classes for the second day in a row.
This was a rare event that had not occurred at the university
since 1999. A blizzard, so I heard, which had washed over the
slate and made the city a true white.

Embolization Two, "stereotactic radiosurgery." It was after
that that the hair started falling away in droves. Months later,
back in the dorm room that I share with Eliza, I started making
a list with everything I knew about hair.

> You cannot tear out hair except from the roots.
> You can cut hair, but you cannot hurt it.

And:

> Mine is black.

Mine is black, but not everyone has black hair. Some have
brown, dirty blonde, platinum blonde, salt-and-pepper. This I
learned in Miss Morris's classroom at the age of seven, when
I moved from Shanghai (where, at the cusp of the twenty-first
century, almost everyone's hair was black) to Dallas. When I
was younger I used to wonder whether people's pubic hair always matched the color on their heads, so I compared notes
with my ginger friend Bryce. I learned a new term that I still
find kind of funny: firecrotch. At the hospital I met a young man
with black hair like mine, except his was wavy and he had dark
skin. Like the other resident physicians, he was startlingly handsome. All of the residents at the Medical Center, if they weren't

already going to be all sorts of -ologists, could have been models. I fantasize about eating in the hospital cafeteria—when I am better and in something other than pastel-colored hospital gowns—and trying to catch the eye of the curly-haired, dark-skinned ER doctor. When I go back to school, I don't go back to the hospital cafeteria. Not because he's seen me use a bedpan. Not because he's already seen my breasts through the size XXL gown (so that I had one less feminine wile at my disposal), the neck hole of which could fit around my waist.

At my parents' house in Texas, the hair loss came to me like an afterthought several days after the radiation. Nothing happened on the first and second days. I washed my hair vigorously.

2-in-1 Shampoo & Conditioner ≠ shampoo + conditioner

Do you know what they do to you when they come at you with numb, technical-sounding words like *stereotactic radiosurgery?* To be sure, it's not serious the way chemo is, so forgive me for adopting a nihilistic attitude. First they have to affix a crown to your head. They do this by drilling four evenly spaced screws into your head, and they stop when you can hear the screw crunching against the cranium, turning bone into fine powder like pestle on mortar. Then they bring in physicists to crunch out the numbers: at what angle to aim the laser, how wide the blast radius should be, how long the exposure.

"This is not the elegant part of the procedure," the resident doctor, who was preparing my head for the placement of the crown, admitted.

I remember saying to her (a nervous young woman with light brown skin, whose dark frames and tightly scrunchied hair kept her from metamorphosing into the heartbreaker that she might have been), to keep the situation light, "Oh, I am so *screwed!*" She laughed but hastily stopped herself.

Post-radiation hair is thinner, fluffier, and paler than pre-radiation hair.

You become deathly afraid that strangers have x-ray vision. And then you grow defiant ("Who gives a shit if everyone can see my bald patch," you'll reason, and then meaningfully leave that knit cap on the bedpost), for there is no room for your dignity to flower into anything except bleary-eyed defiance.

I don't buy a wig. Most of the time a wig serves the purpose of making sure that other people feel comfortable; in my defiance I believed they had no right to that comfort.

Wigmakers are selling not only beauty, but also dignity. But that dignity is fragile.

Later, when the hair has returned, strangers will meet your eyes for the first time. They will say, "Oh, I recognize you." "I've seen you around," they'll say, neglecting to mention that they sat a few rows behind you in Intro to East Asia: Korea and every Monday–Wednesday stared at the back of your yin-yang head.[2]

Hair is best as a binary. You want hair, true, or you want hair, the absence of. Anything in between looks ridiculous, and the comb-over is not a valid option for girls barely out of their teens.

Good friends joke about bad hair days. Acquaintances ask What Happened To questions. Other people look everywhere but at your head.

There is no inherent shame in this, but it is a curious spatial phenomenon. A sociology professor who had nothing to do with me prior to the brain thing approached me one afternoon and said, *It is different for people like you and me, who have seen the other side.*

"Even the Chicago sky doesn't look so gray."

---

2 Radiation's baldness is true baldness, since hair falls out at the roots. No sign of its former inhabitant.

Baldness, at least the involuntary kind, is not "shiny." You get the shininess from repeatedly wearing down your scalp with blades, as though polishing the edge of a kitchen knife, and from moisturizers. Instead, the baldness of radiation is soft and pink—thoroughly unnerving to the touch like eggshells soaked in warm vinegar. What was once alive now pulsates with vengeance at the living.

The truth about the brain thing is that it will always follow me. I don't mean in an abstract way. I mean that every year I cast a die to determine whether I will have another stroke, and there is nothing I can do about it.

Should I write a will? I suppose it would be heartbreaking if it were ever found—heartbreakingly idiotic (like a child's crayon drawing of her nuclear family, the feet sprawled at impossible angles and the hands like flattened cakes of Play-Doh, created with the intent of saving a doomed marriage) if someone found it while I was alive, but plain heartbreaking if someone found it after my death. And anyway I have nothing of value to give anyone except my books—the last vestiges of a brain-rose prematurely pruned.

Another time in the hospital, I had a terrifying nightmare. Half of the world had suffered through a gigantic natural catastrophe of some sort and had to relocate. My family was doing the same; everybody's front lawns were littered with possessions, waiting for their turn to be carried into their new homes. All of my books—compacted into four neat bookshelves—sat there unguarded, tantalizing. And amid the chaos of movers and children and carpenters, some clever soul decided to masquerade as the owner of the books and conduct an impromptu yard sale. In vain I screamed THOSE ARE NOT YOUR BOOKS TO SELL, running from bookshelf to bookshelf, but I succeeded only in garnering jeers. I grew so flustered that my desperation devolved into violence, and I began hitting book buyers left and right. Most of my blows were harmless, but then I distinctly remember picking up *The Great Latke-Hamantash Debate* and jamming the corner into somebody's eye.

This nightmare bothered me for two reasons:

1) Why did the physical manifestation of knowledge matter so much to me? I was not generous with this knowledge, as I should have been—rather, I hoarded earthly paper and ink as though they meant something on their own.

2) Why did I retaliate violently against innocent people, and was my reaction representative of a ruthless survival instinct in waking life? If so, then that makes me morally decrepit. Or at least inauthentic. It must not be a coincidence, then, that the lesion in my brain is next to the pineal gland. Descartes, although he was mistaken about its location, believed it to be the seat of the soul.

In the mornings, clear packing tape clears hairy pillows. The same goes for hairy consciences.

My neurosurgeon is saying things like *You should be able to lead a relatively normal life aside from your unfortunate propensity,* and then he drops the *but.* But the *but* turns out to be surprisingly tame. "But," he says with a winning smile, "stay away from heavy lifting, scuba diving, maybe things that make your blood pump a little too roughly."

I think of asking about orgasms.

I don't.

On another trip to the dream world, I saw my mother. My mother is an illogical woman, as all mothers are prone to be, because her diet consists of far too much true crime. She's forever drawing lines of causation between what I do and what grisly things happen to young girls like myself on a daily basis. In my dream she is clutching fistfuls of her own hair. She is clutching fistfuls of the doctor's hair. The thick strands turn into snakes when they fall to the floor, until the entire office is a writhing, rippling web—the capillaries that I am missing. I can't stand snakes, so I scream and scream.

I remember to call my mother when I wake up. I tell her, *I should be able to lead a relatively normal life, but I should stay away from heavy lifting and scuba diving.* She hears, *I should stay away from anything that makes me too happy or sad or mad.* She also tries to convince me to take the rest of the school year off, a suggestion that I reject.

Your mother cares a lot about your hair, and she will cry over its involuntary separation. That's because she sees the passage of time in the blackness of your hair, and every clump in the trashcan is another reminder that her baby is dying.

"But Mom," you say, "hair is inorganic. It can't *actually* die." Wrong. Hair is organic, because it is made from the protein keratin. You would know that if you studied chemistry like Eliza. Eliza is my roommate, and she is a scientist and a positivist—a bright-eyed, golden-headed student with profound faith in things like reason and logic and the triumph of the superior argument. But neither of us is a very good mathematician, and neither of us believes in predestination. I tell her the truth about the brain thing, my brain thing, and we struggle to calculate the likelihood that I'll survive into my sixties without another stroke. She reacts the same way as I did, her eyes fogging up for a moment but then returning to their original clarity. "I think it's a permutation?"

"Wait a second, I think I have to use integration. We just learned this."

And: "If I do figure out the answer, do you want me to tell you?"

Sure, I say.

Even though half of it is thinner and fluffier than I remember, I have all of my hair now. A bit of it is turning white: a quirk inherited from my father, who started dyeing his hair in his late thirties because at that age it already resembled snow and slate.

But I won't dye. A strand here, a strand there. Like so many reminders to live with authenticity.

FLOYD SKLOOT

Gray Area: Thinking with a Damaged Brain

I USED TO be able to think. My brain's circuits were all connected and I had spark, a quickness of mind that let me function well in the world. There were no problems with numbers or abstract reasoning; I could find the right word, could hold a thought in mind, match faces with names, converse coherently in crowded hallways, learn new tasks. I had a memory and an intuition that I could trust.

All that changed on December 7, 1988, when I contracted a virus that targeted my brain. A decade later, my cane and odd gait are the most visible evidence of damage. But most of the damage is hidden. My cerebral cortex, the gray matter that MIT neuroscientist Steven Pinker likens to "a large sheet of two-dimensional tissue that has been wadded up to fit inside the spherical skull," has been riddled with tiny perforations. This sheet and the thinking it governs are now porous. Invisible to the naked eye but readily seen through brain imaging technology are areas of scar tissue that constrict blood flow. Anatomic holes, the lesions in my gray matter, appear as a scatter of white spots like bubbles or a ghostly pattern of potshots. Their effect is dramatic; I am like the brain-damaged patient described by neuroscientist V. S. Ramachandran in his book *Phantoms in the Brain*: "Parts of her had forever vanished, lost in patches of permanently atrophied brain tissue." More hidden still are lesions in my self, fissures in the thought process that result from this damage to my brain. When the brain changes, the mind changes—these lesions have altered who I am.

"When a disease process hits the brain," writes Dartmouth psychiatry professor Michael Gazzaniga in *Mind Matters*, "the loss of nerve cells is easy to detect." Neurologists have a host of clinical tests that let them observe what a brain-damaged patient can and cannot do. They stroke his sole to test for a spinal reflex known as Babinski's sign or have him stand with feet together and eyes closed to see if the ability to maintain posture is compromised. They ask him to repeat a set of seven random digits forward and four in reverse order, to spell *world* backward, to remember three specific words, such as *barn* and *handsome* and *job*, after a spell of unrelated conversation. A new laboratory technique, positron emission tomography, uses radioactively labeled oxygen or glucose that essentially lights up specific and different areas of the brain being used when a person speaks words or sees words or hears words, revealing the organic location for areas of behavioral malfunction. Another new technique, functional magnetic resonance imaging, allows increases in brain blood flow generated by certain actions to be measured. The resulting computer-generated pictures, eerily colorful relief maps of the brain's lunar topography, pinpoint hidden damage zones.

But I do not need a sophisticated and expensive high-tech test to know what my damaged brain looks like. People living with such injuries know intimately that things are awry. They see it in activities of daily living, in the way simple tasks become unmanageable. This morning, preparing oatmeal for my wife, Beverly, I carefully measured out one-third cup of oats and poured them onto the pan's lid rather than into the bowl. In its absence, a reliably functioning brain is something I can almost feel viscerally. The zip of connection, the shock of axon-to-axon information flow across a synapse, is not simply a textbook affair for me. Sometimes I see my brain as a scalded pudding, with fluky dark spots here and there through its dense layers and small scoops missing. Sometimes I see it as an eviscerated old TV console, wires all disconnected and misconnected, tubes blown, dust in the crevices.

Some of this personal, low-tech evidence is apparent in basic functions like walking, which for me requires intense concentration, as does maintaining balance and even breathing if I am tired. It is apparent in activities requiring the processing of certain fundamental information. For example, no matter how many times I have been shown how to do it, I cannot assemble our vacuum cleaner or our poultry shears or the attachments for our hand-cranked pasta maker. At my writing desk, I finish a note and place the pen in my half-full mug of tea rather than in its holder, which quite obviously teems with other pens. I struggle to figure out how a pillow goes into a pillowcase. I cannot properly adjust Beverly's stereo receiver in order to listen to the radio; it has been and remains useful to me only in its present setting as a CD player. These are all public, easily discernible malfunctions.

However, it is in the utterly private sphere that I most acutely experience how changed I am. Ramachandran compares this to harboring a zombie, playing host to a completely nonconscious being somewhere inside yourself. For me, being brain damaged also has a physical, conscious component. Alone with my ideas and dreams and feelings, turned inward by the isolation and timelessness of chronic illness, I face a kind of ongoing mental vertigo in which thoughts teeter and topple into those fissures of cognition I mentioned earlier. I lose my way. I spend a lot of time staring into space, probably with my jaw dropping, as my concentration fragments and my focus dissolves. Thought itself has become a gray area, a matter of blurred edges and lost distinctions, with little that is sharp about it. This is not the way I used to be.

In their fascinating study *Brain Repair,* an international trio of neuroscientists—Donald G. Stein from America, Simón Brailowsky from Mexico, and Bruno Will from France—report that after injury "both cortical and subcortical structures undergo dramatic changes in the pattern of blood flow and neural activity, even those structures that do not appear to be directly or primarily connected with the zone of injury." From this observation, they conclude that "the entire brain—not just the region around the area of damage—reorganizes in response to brain injury." The implications of this are staggering; my entire brain, the organ by which my very consciousness is controlled, was reorganized one day ten years ago. I went to sleep *here* and woke up *there;* the place looked the same, but nothing in it worked the way it used to.

If Descartes was correct, and to Think is to Be, then what happens when I cannot think, or at least cannot think as I did, cannot think well enough to function in a job or in the world? Who am I?

You should hear me talk. I often come to a complete stop in mid-sentence, unable to find a word I need, and this silence is an apt reflection of the impulse blockage occurring in my brain. Sitting next to Beverly as she drives our pickup truck through Portland traffic at six p.m., I say, "We should have gone for pizza to avoid this blood . . ." and cannot go on. I hear myself; I know I was about to say "blood tower traffic" instead of "rush hour traffic." Or I manifest staggered speech patterns—which feels like speaking with a limp—as I attempt to locate an elusive word. "I went to the . . . *hospital* yesterday for some . . . *tests* because my head . . . *hurt*." Or I blunder on, consumed by a feeling that something is going wrong, as when I put fresh grounds into the empty carafe instead of the filter basket on my coffeemaker, put eye drops in my nose or spray the cleaning mist into my face instead of onto the shower walls. So at the dinner table I might say, "Pass the sawdust" instead of "Pass the rice," knowing even as it happens that I am saying something inappropriate. I might start a conversation about "Winston Salem's new CD" instead of "Wynton Marsalis's" or announce that "the shore is breaking" when I mean to say that "the shower is leaking." There is nothing smooth or unified anymore about the process by which I communicate; it is dis-integrated and unpredictably awkward. My brain has suddenly become like an old man's. Neurologist David Goldblatt has developed a table that correlates cognitive decline in age-associated memory impairment and traumatic brain injury, and the parallels are remarkable. Not gradually, the way such changes occur naturally, but overnight, I was geezered.

It is not just about words. I am also *dyscalculic,* struggling with the math required to halve a recipe or to figure out how many more pages are left in a book I'm reading. If we are on East Eighty-second and Third Avenue in Manhattan, staying with a childhood friend for the week, it is very difficult for me to compute how far away the Gotham Book Mart is over on West Fourty-seventh between Fifth and Sixth, though I spent much of my childhood in the city.

Because it is a place where I still try to operate normally, the kitchen is an ideal neurological observatory. After putting the leftover chicken in a plastic bag, I stick it back in the oven instead of the refrigerator. I put the freshly cleaned pan in the refrigerator, which is how I figure out that I must have put the chicken someplace else because it's missing. I pick up a chef's knife by its blade. I cut off an eighth of a giant white onion and then try to stuff the remainder into a recycled sixteen-ounce yo-gurt container that might just hold the small portion I set aside. I assemble ingredients for a vinaigrette dressing, pouring the oil into an old olive jar, adding balsamic vinegar, mustard, a touch of fresh lemon juice and spices. Then I screw the lid on upside-down and shake vigorously, spewing the contents everywhere. I stack the newspaper in the woodstove for recycling. I walk the garbage up our two-hundred-yard-long driveway and try to put it in the mailbox instead of the trash container.

At home is one thing; when I perform these gaffes in public, the effect is often humiliating. I can be a spectacle. In a music store last fall, I was seeking an instruction book for Beverly, who wanted to relearn how to play her old recorder. She informed me that there were several kinds of recorders; it was important to buy exactly the right category of book since instructions for a soprano recorder would do her no good while learning on an alto. I made my way up to the counter and nodded when the saleswoman asked what I wanted. Nothing came out of my mouth, but I did manage to gesture over my right shoulder like an umpire signaling an out. I knew I was in trouble, but forged ahead anyway, saying, "Where are the books for som-brero reporters?" Last summer in Manhattan, I routinely exited the subway stations and led Beverly in the wrong direction, no matter which way we intended to go. She kept saying things like "I think west is *that* way, sweetie," while I confidently and mistakenly headed east, into the glare of the morning sun, or "Isn't that the river?" as I led her away from our riverside destination. Last week, in downtown Portland on a warm November morning, I stopped at the corner of Tenth and Burnside, one of the busiest crossings in the city, carefully checked the traffic light (red) and the traffic lanes (bus coming), and started to walk into the street. A muttering transient standing beside me on his way to Powell's Books, where he was going to trade in his overnight haul of tomes for cash, grabbed my shoulder just in time.

At home or not at home, it ultimately makes no difference. The sensation of *dysfunctional mentation* is like being caught in a spiral of lostness. Outside the house, I operate with sporadic success, often not knowing where I am or where I'm going or what I'm doing. Inside the house, the same feelings often apply and I find myself standing on the top of the staircase wondering why I am going down. Even inside my head there is a feeling of being lost, thoughts that go nowhere, emptiness where I expect to find words or ideas, dreams I never remember.

Back in the fall, when it was Beverly's birthday, at least I did remember to go to the music store. More often, I forget what I am after within seconds of beginning the search. As she gets dressed for work, Beverly will tell me what she wants packed for lunch and I will forget her menu by the time I get up the fourteen stairs. Now I write her order down like a waiter. Sometimes I think I should carry a pen and paper at all times. In the midst of preparing a salad, I stop to walk the four paces over to the little desk where we keep our shopping list and forget "tomatoes" by the time I get there. Between looking up a phone number and dialing it, I forget the sequence. I need the whole phone book on my speed-dial system.

Though they appear without warning, these snafus are no longer strange to me. I know where they come from. As Dr. Richard M. Restak notes in *The Modular Brain,* "A common error frequently resulting from brain damage involves producing a semantically related word instead of the correct response." But these paraphasias and neologisms, my *expressive aphasias,* and my dyscalculas and my failures to process—the rapids of confusion through which I feel myself flailing—though common for me and others with brain damage, are more than symptoms to me. They are also more than what neurologists like to call *deficits,* the word of choice when describing impairment or incapacity of neurological function, as Oliver Sacks explains in his introduction to *The Man Who Mistook His Wife for a Hat.* These "deficits" have been incorporated into my very being, my consciousness. They are now part of my repertoire. Deficits imply losses; I have to know how to see them as gains.

Practitioners of neuroscience call the damage caused by trauma, stroke, or disease "an insult to the brain." So pervasive is this language that the states of Georgia, Kentucky, and Minnesota, among others, incorporate the phrase "insult to the brain" in their statutory definitions of traumatic brain injury for disability determinations. Such insults, according to the Brain Injury Association of Utah, "may produce a diminished or altered state of consciousness, which results in an impairment of cognitive abilities or physical functioning." The language used is so cool. There is this sentence from the website NeuroAdvance.com: "When there is an insult to the brain, some of the cells die." Yes.

Insult is an exquisitely zany word for the catastrophic neurological event it is meant to describe. In current usage, of course, it generally refers to an offensive remark or action, an affront, a violation of mannerly conduct. To insult is to treat with gross insensitivity, insolence, or contemptuous rudeness. The medical meaning, however, as with so many other medical words and phrases, is different, older, linked to a sense of the word that is some two or three centuries out of date. *Insult* comes from the Latin compound verb *insultare,* which means "to jump on" and is also the root word for *assault* and *assail.* It's a word that connotes aggressive physical abuse, an attack. Originally, it suggested leaping upon the prostrate body of a foe, which may be how its link to contemptuous action was forged.

Though "an insult to the brain" (a blow to the head, a metal shard through the skull, a stroke, a viral "attack") is a kind of assault, I am curious about the way *contempt* has found its way into the matter. Contempt was always part of the meaning of insult and now it is primary to the meaning. Certainly a virus is not acting contemptuously when it targets the brain; neither is the pavement nor steering wheel nor falling wrench nor clot of blood nor most other agents of "insult." But I think society at large, medical scientists, insurers, legislators, and the person on the street do feel a kind of contempt for the brain-damaged with their comical way of walking, their odd patterns of speech or ways in which neurological damage is expressed, their apparent stupidity, their abnormality. The damage done to a brain seems to evoke disdain in those who observe it and shame or disgrace in those who experience it.

Poet Peter Davison has noticed the resonant irony of the phrase "an insult to the brain" and made use of it in his poem "The Obituary Writer." Thinking about the suicide of John Berryman, the heavily addicted poet whose long-expected death in 1972 followed years of public behavior symptomatic of brain damage, Davison writes that "his hullabaloos / of falling-down drunkenness were an insult to the brain." In this poem, toying with the meaning of the phrase, Davison suggests that Berryman's drinking may have been an insult to his brain, technically speaking, but that watching him was, for a friend, another kind of brain insult. He has grasped the fatuousness of the phrase as a medical term, its inherent judgment of contempt, and made use of it for its poetic ambiguity.

But I have become enamored of the idea that my brain has been insulted by a virus. I use it as motivation. There is a long tradition of avenging insults through duels or counterinsults, through litigation, through the public humiliation of the original insult. So I write. I avenge myself on an insult that was meant, it feels, to silence me by compromising my word-finding capacity, my ability to concentrate and remember, to spell or conceptualize, to express myself, to think.

The duel is fought over and over. I have developed certain habits that enable me to work—a team of seconds, to elaborate this metaphor of a duel. I must be willing to write slowly, to skip or leave blank spaces where I cannot find words that I seek, to compose in fragments and without an overall ordering principle or imposed form. I explore and make discoveries in my writing now, never quite sure where I am going, but willing to let things ride and discover later how they all fit together. Every time I finish an essay or poem or piece of fiction, it feels as though I have faced down the insult.

In his book *Creating Mind*, Harvard neurobiologist John E. Dowling says, "The cerebral cortex of the human brain, the seat of higher neural function—perception, memory, language, and intelligence—is far more developed than is the cerebral cortex of any other vertebrate." Our gray matter is what makes us human. Dowling goes on to say that "because of the added neural cells and cortical development in the human brain, new facets of mind emerge." Like the fractured facet of a gemstone or crystal, like a crack in the facet of a bone, a chipped facet of mind corrupts the whole, and this is what an insult to the brain attacks.

Though people long believed, with Aristotle, that the mind was located within the heart, the link between brain and mind is by now a basic fact of cognitive science. Like countless others, I am living proof of it. Indeed, it is by studying the behavior of brain-damaged patients like me that medical science first learned, for example, that the brain is modular, with specific areas responsible for specific functions, or that functions on one side of the body are controlled by areas on the opposite side of the brain. "The odd behavior of these patients," says Ramachandran, speaking of the brain-damaged, "can help us solve the mystery of how various parts of the brain create a useful representation of the external world and generate the illusion of 'self' that endures in space and time." Unfortunately, there is ample opportunity to observe this in action since, according to the Brain Injury Association, more than two million Americans suffer traumatic brain injury every year, a total that does not include damage by disease.

"Change the brain, change the person," says Restak in *The Modular Brain*. But how, exactly? No one has yet explained the way a brain produces what we think of as consciousness. How does the firing of electrical impulse across a synapse produce love, math, nightmare, theology, appetite? Stated more traditionally, how do brain and mind interact? Bookstore shelves are now filled with books, like Steven Pinker's brilliant 1997 study *How the Mind Works,* that attempt to explain how a three-pound organ that is the consistency of Jell-O makes us see, think, feel, choose, and act. "The mind is not the brain," Pinker says, "but what the brain does."

And what the brain does, according to Pinker, "is information processing, or computation." We think we think with our brain. But in doing its job of creating consciousness, the brain actually relies upon a vast network of systems and is connected to everything—eyes, ears, skin, limbs, nerves. As Dowling so dourly puts it, our mental function, our mind—memory, feelings, emotions, awareness, understanding, creativity—"is an emergent property of brain function." In other words: "What we refer to as mind is a natural consequence of complex and higher neural processing."

The key word is *processing*. We actually think with our whole body. The brain, however, takes what is shipped to it, crunches the data, and sends back instructions. It converts; it generates results. Or, when damaged, does not. There is nothing wrong with my sensory receptors, for instance. I see quite well. I can hear and smell. My speech mechanisms (tongue, lips, nerves) are intact. My skin remains sensitive. But it's in putting things together that I fail. Messages get garbled, blocked, missed. There is, it sometimes seems, a lot of static when I try to think, and this is the gray area where nothing is clear any longer.

Neurons, the brain's nerve cells, are designed to process information. They "receive, integrate and transmit," as Dowling says, receiving input from dendrites and transmitting output along axons, sending messages to one another across chemical passages called synapses. When there are lesions like the ones that riddle my gray matter, processing is compromised. Not only that, certain cells have simply died and with them the receiving, integrating, and transmitting functions they performed.

My mind does not make connections because, in essence, some of my brain's connectors have been broken or frayed. I simply have less to work with and it is no surprise that my IQ dropped measurably in the aftermath of my illness. Failing to make connections, on both the physical and metaphysical levels, is distressing. It is very difficult for me to free-associate; my stream of consciousness does not absorb runoff or feeder streams well, but rushes headlong instead. Mental activity that should follow a distinct pattern does not and, indeed, I experience my thought process as subject to random misfirings. I do not feel in control of my intelligence. Saying, "Pass me the tracks" when I intended to say, "Pass me the gravy" is a nifty example. Was it because *gravy* sounds like *grooves,* which led to *tracks,* or because my tendency to spill gravy leaves tracks on my clothes? A misfire, a glitch in the gray area that thought has become for me, and as a result my ability to express myself is compromised. My very nature seems to have altered.

I am also easily overloaded. I cannot read the menu or converse in a crowded, noisy restaurant. I get exhausted at Portland Trailblazers basketball games, with all the visual and aural imagery, all the manufactured commotion, so I stopped going nine years ago. My hands are scarred from burns and cuts that occurred when I tried to cook and converse at the same time. I cannot drive in traffic, especially in our standard transmission pickup truck. I cannot talk about, say, the fiction of Thomas Hardy while I drive; I need to be given directions in small doses rather than all at once, and need those directions to be given precisely at the time I must make the required turn. This is, as

Restak explains, because driving and talking about Hardy, or driving and processing information about where to turn, are handled by different parts of the brain and my brain's parts have trouble working together. I used to write accompanied by soft jazz, but now the least pattern of noises distracts me and shatters concentration. My entire writing process, in fact, has been transformed as I learned to work with my newly configured brain and its strange snags. I have become an avid note taker, a jotter of random thoughts that might or might not find their way together or amount to anything, a writer of bursts instead of steady work. A slight interruption—the movement of my cat across my window view, the call of a hawk, a spell of coughing—will not just make me lose my train of thought, it will leave me at the station for the rest of the day.

I have just finished reading a new book about Muhammad Ali, *King of the World*, by David Remnick. I anticipated identifying a bit with Ali, now suffering from Parkinson's disease, who shows so strikingly what brain damage can do, stripped as he is of so many of the functions—speech, movement, spontaneity—that once characterized him. But it was reading about Floyd Patterson that got me.

Patterson was a childhood hero of mine. Not only did we share a rare first name, we lived in neighboring towns—he was in Rockville Center, on Long Island, while I was five minutes away in Long Beach, just across the bridge. I was nine when he beat Archie Moore to take the heavyweight championship belt, almost twelve when he lost it to Ingemar Johannson, and almost thirteen when he so memorably won it back. The image of Johannson's left leg quivering as he lay unconscious on the mat is one of those vivid memories that endure (because, apparently, it is stored in a different part of the brain than other, less momentous memories). That Floyd, like me, was small of stature in his world, was shy and vulnerable, and I was powerfully drawn to him.

During his sixty-four professional fights, his long amateur

career, his many rounds of sparring to prepare for fights, Patterson absorbed a tremendous amount of damage to his brain. He's now in his sixties, and his ability to think is devastated. Testifying in court earlier this year in his capacity as head of the New York State Athletic Commission, Patterson "generally seemed lost," according to Remnick. He could not remember the names of his fellow commissioners, his phone number or secretary's name or lawyer's name. He could not remember the year of his greatest fight, against Archie Moore, or "the most basic rules of boxing (the size of the ring, the number of rounds in a championship fight)." He kept responding to questions by saying, "It's hard to think when I'm tired."

Finally, admitting "I'm lost," he said, "Sometimes I can't even remember my wife's name, and I've been married thirty-two, thirty-three years." He added again that it was hard for him to think when he was tired. "Sometimes, I can't even remember my own name."

People often ask if I will ever "get better." In part, I think what they wonder about is whether the brain can heal itself. Will I be able, either suddenly or gradually, to think as I once did? Will I toss aside the cane, be free of symptoms, have all the functions governed by my brain restored to smooth service, rejoin the world of work and long-distance running? The question tends to catch me by surprise because I believe I have stopped asking it myself.

The conventional wisdom has long been that brains do not repair themselves. Other body tissue, other kinds of cells, are replaced after damage, but "when brain cells are lost because of injury or disease," Dowling wrote as recently as 1998, "they are not replaced." We have, he says, as many brain cells at age one as we will ever have. This has been a fundamental tenet of neuroscience, yet it has also long been clear that people do recover—fully or in part—from brain injury. Some stroke victims relearn to walk and talk; feeling returns in once numbed limbs. Children—especially children—recover and show no lasting ill effects from catastrophic injuries or coma-inducing bouts of meningitis.

So brain cells do not get replaced or repaired, but brain-damaged people occasionally do regain function. In a sense, then, the brain heals, but its cells do not.

In *Confronting Traumatic Brain Injury*, Texas bioethicist William J. Winslade says, "Scientists still don't understand how the brain heals itself." He adds that although "until recently, neuroscientists thought that much of the loss of capabilities due to brain damage was irreversible," patients recover spontaneously and rehabilitation programs "can restore cognitive and functional skills and emotional and experiential capacity, at least in part."

There are in general five theories about the way people might recover function lost to brain damage. One suggests that we do not need all our brain because we use only a small part of it to function. Another is that some brain tissue can be made to take over functions lost to damage elsewhere. Connected to this is the idea that the brain has a backup mechanism in place allowing cells to take over like understudies. Rehabilitation can teach people new ways to perform some old tasks, bypassing the whole damaged area altogether. And finally, there is the theory that in time, and after the chemical shock of the original injury, things return to normal and we just get better.

It is probably true that, for me, a few of these healing phenomena have taken place. I have, for instance, gotten more adept at tying my shoes, taking a shower, driving for short periods. With careful preparation, I can appear in public to read from my work or attend a party. I have developed techniques to slow down my interactions with people or to incorporate my mistakes into a longer-term process of communication or composition. I may not be very good in spontaneous situations, but given time to craft my responses, I can sometimes do well. But I still can't think.

A recent development promises to up the ante in the game of recovery from brain damage. The *New York Times* reported in October of 1998 that "adult humans can generate new brain cells." A team at the Salk Institute for Biological Studies in La Jolla, California, observed new growth in cells of the hippocampus, which controls learning and memory in the brain. The team's leader, Dr. Fred Gage, expressed the usual cautions: more time is needed to "learn whether new cell creation can be put to work" and under what conditions. But the findings were deemed both "interesting" and "important."

There is only one sensible response to news like this. It has no personal meaning to me. Clinical use of the finding lies so far in the future as to be useless, even if regenerating cells could restore my lost functions. Best not to think about this sort of thing.

Because, in fact, the question of whether I will ever get better is meaningless. To continue looking outside for a cure, a "magic bullet," some combination of therapies and treatments and chemicals to restore what I have lost, is to miss the point altogether. Certainly if a safe, effective way existed to resurrect dead cells, or generate replacements, and if this somehow guaranteed that I would flash back or flash forward to "be the person I was," it would be tempting to try.

But how would that be? Would the memories that have vanished reappear? Not likely. Would I be like the man, blind for decades, who had sight restored and could not handle the experience of vision, could not make sense of a world he could see? I am, in fact, who I am now. I have changed. I have learned to live and live richly as I am now. Slowed down, softer, more heedful of all that I see and hear and feel, more removed from the hubbub, more internal. I have made certain decisions, such as moving from the city to a remote rural hilltop in the middle of acres of forest, that have turned out to be good for my health and even my soul. I have gained the love of a woman who knew me before I got sick and likes me much better now. Certainly I want to be well. I miss being able to think clearly and sharply, to function in the world, to move with grace. I miss the feeling of coherence or integrity that comes with a functional brain. I feel old before my time.

In many important respects, then, I have already gotten better. I continue to learn new ways of living with a damaged brain. I continue to make progress, to avenge the insult, to see my way around the gray area. But no, I am not going to be the man I was. In this, I am hardly alone.

DINTY W. MOORE

The Aquatic Ape Hypothesis,
or How I Learned to Love My Paranasal Sinuses

UNTIL JUST A few weeks ago, here is everything I knew about my sinuses:

1. They are inside my head.
2. They are usually clogged with horrible mucus.
3. The horrible mucus leaks out of my nostrils.
4. Sinuses are disgusting, and the less time spent thinking about them the better.

Or so I thought.

It turns out that modern medicine is mind-blowing, and I mean that in a thoroughly positive way. I might have meant it otherwise had my doctor's hand somehow slipped during surgery, but that's getting well ahead of the story.

For now, here's what you need to know:

After fifty years of inadequate breathing, decades of pulsing discomfort, a general sense of "I hate my sinuses, why do I even have them," I was informed by modern medicine, in the form of a young, slender, oddly confident ENT specialist, that my problem was not my sinuses *per se,* but sinus polyps—grape-sized blobs of I-don't-know-and-I-didn't-ask.

These grape-sized blobs of I-don't-know-and-I-didn't-ask are what kept my sinuses from filling with air. They also kept them from flushing out all the horrible mucus. Thus: infection, pain, poor breathing, infection, gunk, embarrassment, infection, more pain, a box of Kleenex on every flat surface of my home, burning, swelling, infection, pain. Repeat cycle once each month.

Then modern medicine suggested: "We can clear those out."

"How?" I asked.

"Well, we go up through the nostrils . . ." the doctor said.

"The nostrils, you say?"

"Yes," the young physician answered, and then he offered a sentence that contained the word "scraping," and I removed myself from all conscious comprehension for about ten seconds, until he said, "Of course, we wouldn't want to scrape too much, because the bone separating your sinuses from your brain is very thin."

As I said: Potentially mind-blowing.

It was at that juncture that I stopped listening for about thirty seconds, until the doctor added, "So we should probably schedule this up in Columbus, just to be on the safe side."

I remember wondering why the thin layer of bone separating my sinuses from my brain would be less likely to perforate catastrophically in Columbus, the capital of Ohio, about eighty miles upstream from the small college town where my sinuses usually clog themselves. But it didn't take long before the doctor said, "Imaging."

"Oh," I nodded, trying to look respectful and informed. "Who's Imogene?"

So, here are six *actual* facts I didn't know about my sinuses before Doctor Gallant (not his real name, but it should be) entered the picture:

1. There are not two but four sinus cavities in the skull—one on either side of the nose, but also one above each eye, behind the eyebrow.

2. Scientists can't agree why these openings exist.

3. One theory is that they decrease the weight of the skull, making it easier to hold up our heads all day.

4. Another theory is that they act as shock absorbers, decreasing injury when the head hits something harder than a pillow.

5. The goop we all despise exists for good reason: to capture viruses, bacteria, and other airborne particles before they reach our lungs.

6. When we are sick, mucus production can increase to two

liters a day. Think two-liter Pepsi bottle, and then get entirely grossed out.

There was, it turns out, no Imogene.

Dr. Gallant scheduled me in early August for Computer-Assisted Endoscopic Sinus Surgery. This involved the insertion of a very thin, fiber-optic scope into my nose and the use of micro instruments (aka "scrapers") to remove the little grape-sized blobs of I-didn't-ask. Of course, if the doctor was going to wander around with tiny X-ACTO knives, it would be good for him to see where he was scraping. The hospital in Columbus, it turns out, had *imaging* technology.

First, though, I had to get medically cleared for the operation. Because I am in advanced middle age, I have many doctors; we humans accumulate them like barnacles attached to an aging frigate. None of my many doctors, of course, could figure out how to share information with any of my other many doctors, including doctors whose offices are one floor apart in the same medical complex. "I can just walk it down," I would say, but they had protocols, and costly computer systems that couldn't talk to one another, or do anything really, except billing.

The billing always worked.

Nonetheless, August rolled around, and I presented myself at the Outpatient Surgery Center, located just a few blocks from the enormous university teaching hospital, and all was well, except at the last minute I mentioned that I'd recently been diagnosed with sleep apnea, a Greek word that allows doctors to bill you at two hundred times the rate they might if we just called it snoring.

My procedure was delayed while the medical team endeavored to learn my sleep apnea score, which somehow had never found its way into any of my voluminous medical records.

"I believe I scored well," I said, which didn't satisfy the anesthesiologist's curiosity at all.

Sixteen computers in sixteen different medical offices spread across most of southern Ohio refused to speak to one another for a good bit of the morning, until the resourceful anesthesiologist finally just picked up his cell phone and dialed.

The last voice I heard before succumbing to the happy gas was the masked cell phone user reacting to the score he was given:

"Holy cow!"

I assume the doctor has wonderful memories of touring the folds and caverns behind my facial bones, but since Gallant and his team kept me sedated and oblivious, my only way of describing what occurred is to watch similar procedures on YouTube, where, it turns out, hundreds of doctors have recorded thousands of excruciating hours of footage revealing just about any medical technique you might want to watch. It is creepy, to be honest, because the doctors in these videos talk animatedly at the camera for most of the operation, and I keep wanting to shout, "Oh my God, focus on the patient, focus on the patient!"

The online videos of Computer-Assisted Endoscopic Sinus Surgery using image guidance aren't pretty, believe me. The flexible tube inserted through the nostril contains both a light source and a camera, and the inner walls, gooey corners, and grape-sized I-don't-know-whats are revealed on a TV monitor. The videos look like outtakes from a movie titled *Journey to the Center of an Astonishingly Gross Earth,* or perhaps extremely poor-quality porn, shot *way* too close up.

I awoke from my procedure feeling quite chipper. Until Dr. Gallant and the anesthesiologist informed me I would not be heading home as planned, but staying the night in a local hospital. My "holy cow!" sleep apnea score, they concluded, combined with the amount of anesthesia it took to knock me out for surgery, risked that unpleasant moment where my airwaves would briefly shut off breathing, and my reflexes would just roll over and say, "Oh don't wake us now, we're having such a nice dream."

In other words, I would asphyxiate.

The medical chaps, as they loved to say over and over again, decided to "exercise a little extra caution."

This did not sit well with me. I wanted to recover at home, as "outpatient" surgery suggested, both because of sentimental reasons, but also because I had planned my "at home" outpatient recovery in exquisite detail, a sort of one-man New Year's Eve celebration featuring cold beer, junk television, nose bandages, and pain killers. What could go wrong?

I wasn't going to find out because I wasn't going home, which was bad enough. Worse yet was when the hospital reported having no open rooms.

The real problem was that I felt absolutely fine. Anesthesia has the odd effect of energizing me immediately after awakening, rather than leaving me drowsy, but given my "post-op" status, I was stuck with two choices—either lie on my back and complain, or sit up just a little, sip water, and complain.

Three hours of this, until finally I was cleared for a room in the hospital six blocks away, and then—yes, only then—a nurse informs me that an ambulance has been called, and that will take ". . . about three more hours."

"Your case is not urgent," she added.

What I said in response may not have been polite, and I hereby apologize to anyone anywhere who has ever worked in the medical care profession.

About this point, I went to work trying to convince the nursing staff that I easily enough could *walk* the six blocks to the hospital. Or I could drive, if they lent me a car. Or one of them could drive me, and I'd buy ice cream on the way.

Miraculously, and to the boundless relief of the nurses, my ambulance arrived a full hour and a half early, and I was quickly strapped in, attached to four thousand wires, monitoring every inch of my body except perhaps my nose, where I believe the surgery had been performed. And then, finally, I was driven the three-minutes' distance from the surgery center to the medical center, at about twenty miles per hour, no lights, no siren.

At one point, concerned that her patient might be disoriented by this wild ride, the med tech in the back asked me the name of the current president.

"Sarah Palin," I answered, hoping to exhibit the fine nuance of my post-operative intellectual irony.

"Ha!" she answered with no hint of humor. "Don't we wish."

Faster than one can say Affordable Care Act, I was whisked into my room, on the hospital's fifth floor. The man in the bed across from me was glad for company, because he had quite the story to tell, one I heard about eight times in the next four hours.

Mr. Deeter was from Akron, and his job, he told me, was to service giant transformers, the ones you see along the roadside surrounded by ten-foot cyclone fencing with signs reading: "High Voltage! Do Not Enter!"

Mr. Deeter regularly ignored those signs—it was, in fact, his job to do so. That morning he had been pulling oil from the engine of one of these powerful transformers—"with a syringe," he shouted across the two beds, "the way a nurse takes blood"—when his bare arm touched something it should not have touched, and 81,000 volts of electricity coursed through his body.

"I let out a yelp," he told me. "And BAM! Next thing I knew I was knocked back up against the fence."

He stopped for a moment, studied my face. What he saw was an expression that best translates as "And you *lived*?"

Mr. Deeter seemed to be rounding sixty or so, with a short, military haircut, the fit physique of a man who works outside with tools, and a deep, no-nonsense voice. He was proud of his ability to survive the massive burst of voltage, or maybe he was still in shock. Either way, he repeated his story to everyone who entered the room.

"Couldn't feel my arm at first, so I looked down, and, yup, it was still attached." He would pause here for effect. "Then I went back to work, siphoning out the oil. I noticed this burn on my elbow, and thought, 'Oh nuts! I guess I should call this one in.' But I didn't."

Turned out Mr. Deeter had two small, round burns: one on his elbow, just an inch or so from where his safety gloves ended, and one on his chest, where the voltage apparently surged back out of his body.

He didn't call to report the accident until a co-worker showed up, and said, "Deeter, you don't look so good."

"He was right. I called it in. Now I'm here."

He didn't look like a man shot through with electricity. He looked fine, as fine as I felt. He also looked trapped, like he'd rather be anywhere, even back servicing generators, than in that hospital room.

I knew exactly how he felt.

Scientists, as I said earlier, can't agree on why we have sinuses.

The make-our-heads-lighter-so-we-can-hold-them-erect notion has its staunch advocates, as does the shock-absorber-in-the-skull idea, but, hands-down, my favorite theory posits that we—you, me, Mr. Deeter, and Sarah Palin alike—are descended from aquatic apes.

The theory goes like this: a group of prehistoric primates, cleverer than most, noticed that river banks and sea shores produced much better food than did arid grasslands, so they descended from their treetops and acquired waterfront property.

Over time, through the exquisite magic of evolution, these apes evolved an upright stance, allowing them to stand in the water and freeing up their hands to crack shellfish. Eventually they also lost their body hair, developing instead a thick layer of subcutaneous fat (to keep warm in the water). They learned to swim.

And this, if you believe Peter Rhys Evans, a British expert on head-and-neck physiology, also explains our sinus cavities.

Compared to other primates, humans have particularly large openings in the skull, Rhys Evans notes. "It makes no sense until we consider the evolutionary perspective. Then it becomes clear: our sinuses acted as buoyancy aids that helped keep our heads above water."

He adds further evidence: unlike our ape cousins, humans have an unusually strong diving reflex, a unique nose shape that shields our nostrils when we dive below the surface, and partial webbing between our fingers and toes.

Not all scientists agree, because if they did, how could they write hundreds of scholarly articles arguing over every detail— but a good many do agree. And who doesn't like a spirited squabble over how primeval monkeys transformed themselves over time into twenty-first-century hipsters wearing skinny jeans and taking selfies?

Turns out, it all started at the oyster bar.

Why exactly do human beings have unique tongue prints?

Why do we have that vertical groove on the surface of our upper lip?

What's the meaning of goosebumps?

What purpose does the uvula serve, and why does it sound so dirty?

If Mr. Deeter could absorb thousands of volts of electricity through his arm and shoot it back out of his chest, sustaining little more than a few surface burns, and then go back to work for thirty minutes before deciding to call his supervisor, why can't monkeys evolve large open spaces in their skulls to keep their heads above water as they float down the lazy river, popping tasty minnows into their hungry mouths?

I'm talking about the glorious mystery of the body here, which might sound like a pickup line, but I don't mean it that way.

Goosebumps, by the way, occur when tiny muscles around the base of each hair tense, pulling the hair more erect. Back when we were apes, our fur would stand on end, to make us look larger, scarier, more powerful. Now, we just look silly.

Our bodies, even our sinuses, are simply miraculous. I've progressed from hating my goopy head cavities to being damned proud of them.

They exist for a reason.

A good reason.

They exist because somehow, somewhere in time, an ape looked around and thought, "Man, you know what I could go for right now? Shrimp cocktail."

DANIELLE R. SPENCER

Looking Back

I AM SUSPENDED from his large brown eye. It is concentrating, peering closely into my own eye. I am afraid I will fall. I feel the heat of the lamp on my forehead, this surgeon's hands on my face, and the sharp pressure of metal on my eye surfacing through the numbing drops. I feel a single trickle of sweat down my back but not the chair beneath me—and through the shifting and dissolving blue shapes there is nothing tethering me in place except this eye looking down into mine.

I am hoping this surgeon can straighten my wayward eye. For as long as I can remember I have seen my crossed eyes reflected back in the way other people look at me: the flicker of recognition that something is off; then looking away; then back, but differently, now. If I look at you directly, I'll see you seeing how I look, so it's best if I just look away. When I see someone with this condition—strabismus—I blink with recognition at the furtive head posture, the particular elliptical ways we have of looking.

Seeing the world this way is like thinking in simile or metaphor. Both are indirect, roundabout, oblique. I can't separate the way I see and think from my crossed eyes. An exacting French teacher once crumpled my essay up in her hands and placed it on the table in front of me. This is how you are writing, she explained, pointing. I looked down at the torted ball of a page and saw that she was right; that sight and language and mind were bound together, twisting inward.

In my baby pictures I am round and happy, a laughing little girl in a striped bonnet with plump cheeks and both eyes tucked up and inward toward my nose. As a young child, I had two surgeries, one on each eye, which left one straight and the other pointing in. Without proper alignment, I didn't have typical depth-perception—no stereopsis—though to me the world looked the way it always had. In first grade, classmates began to cross their own eyes back at me, laughing. When I came home from this teasing, my parents explained that I was fortunate to have my sight, and that we play the cards we're dealt, that having misaligned eyes is only superficial. I accepted this explanation, and as I grew older I was ashamed to care about something so trivial and so external.

But in fact it's not trivial. It's not external. Eyes are the nexus of perception and identity—seeing and being seen—and a crooked gaze shorts that circuitry, its repellence unfurling in language. In English, *cockeyed* describes anything that's absurd or askew; in Hungarian *bandzsa* also means "stupid"; in Mandarin (对眼) means "easily enraged." In German (*schielend*) and Hebrew (פּוֹזֵל) the phrase doubles as "envious" or "covetous," and the Swedish *skelögd* also means "distractible." In French, the term *louche* also means "immoral," "shifty," "disreputable"; as one eye looks toward you while the other looks away, it describes anything that is ambiguous or untrustworthy.

Historically, misaligned eyes were thought to be a bad omen. In English and American folklore, you are supposed to bow or spit when you meet a cross-eyed person, and cross the street to avoid misfortune. To this day, studies show that people with strabismus are perceived to be less reliable, intelligent, honest, and employable than people with straight eyes. It's the "Cross-Eyed Blues": *Folks who's got them cross eyes, things they see is always wrong. That's why me and cross eyes never gonna get along.* That vision should be so packed with symbolism and associations isn't so surprising. The eyes are the gatekeepers between the mind and the body, between ourselves and the world.

On the other side of the spectrum is the view that the mind and perceptions are separate and distinct. René Descartes, father of the Enlightenment, described sight as geometry, abstracting it from the motions and vagaries of incarnate experience. Here, the disembodied eye watches, steadily and objectively, from a fixed point. Yet his great epiphany about the power of the rational mind arrived, in fact, when he contemplated why he'd always been attracted to cross-eyed women—and realized that it was because of a childhood crush on a strabismic girl. After he reflected and recognized that being cross-eyed was a flaw, he shed the "illogical" attraction, and, thus, reason won out over the passions. And so the frozen stigma of misaligned eyes appears, even here—caught in the amber of the mind's triumph over the body.

Strabismus most often originates in the mind: the neurological wiring that keeps the eyes aligned doesn't function properly. But the treatment is a mechanical fix of the body: typically moving one or more of the muscles attached to the eye—forward to increase the torque, or back to weaken it. So I, with my crossed eyes, decided to exercise my rationality and give surgery another try in adulthood. The procedure is considered functional, not cosmetic, and the odds favor a good result. Even so, I felt like I was betraying myself and the moral code of my childhood in wanting to change my eyes. To be sure, there are far worse burdens to bear. Too many even to begin counting them. But what if this one didn't have to be borne, after all?

I first went to a surgeon near home who assured me he could straighten my eyes. Bustling, short and round, a chattering magpie, he kept finishing my sentences for me—but not with the right endings, not with the words I'd meant to use. It was as though he could make me as he thought I should be just by saying so, but he couldn't. He operated once, and then again, and left me with my left eye turned not in but out, slipping farther and farther toward my ear.

And so last fall I took the train up north to consult a specialist—this surgeon with the large brown eyes. His office was in a renowned pediatric teaching hospital, as strabismus is most often treated in children. I wound through a maze of halls painted in primary colors, past bald little boys staring at goldfish circling around a vast tank. In the oblong ophthalmology exam rooms, stuffed cows looked down equally from above the eye charts. The surgeon entered and greeted me, tall and thoughtful, with a clutch of observers following behind. He folded himself into a chair, looked and listened intently, a gentle humor weaving through his precision. I trusted him instinctively, and felt encouraged.

His exam uncovered a puzzling surprise, though. When he held up fingers to my left and right—something my hometown eye surgeon hadn't checked—I couldn't count the ones on the right. He handed me orders for an MRI, told me not to wait to have it done. I took the train back home and called my internist, someone I thought of as eternally bemused and unflappable. When I described the visual field abnormalities, though, his voice shifted register: Call these specialists, he told me. Go see them. I'd been so accustomed to his wry stoicism that it took me a few moments to understand that he was giving me instructions. Yes, that's what I'm saying, that this is what I want you to do, he had to explain.

By now I was starting to understand: disruptions of the visual pathways can indicate problems that range from mild, such as migraines, to terrifying—such as a brain tumor. Suddenly the world was shifted from where it was supposed to be, refracted by the prism of fear: fear of what this could be, fear of not knowing.

So off I tumbled to a gaggle of neurologists, who watched me keenly and asked a lot of questions. First came the neuro-ophthalmologist, who held my visual field charts in one hand and my bicycle helmet in the other—they were the same semicircular, bread-loaf shape—and looked back and forth between the two incredulously, shaking them for emphasis: This is *your* helmet? And these are *your* visual fields? You rode your bicycle here *today*? Through midtown *traffic*? Let's scan your brain, shall we? Next was the neurologist, who ran his hands lightly over my arms and legs, chatting as he administered his odd motor tests, like cocktail party conversation mixed with a high-stakes game of Simon Says: So, you're an art director? Close your eyes and touch your finger to your nose. So, you run in the mornings? Lie down and run your heel up the front of your opposite shin. Finally came the neurosurgeon—cheerful, buck-toothed, and disheveled. He sat on the examining table in a rumpled white coat with his legs sticking out straight over the side, jabbed his pen at a small, mottled-gray plastic model of a brain with a pharmaceutical logo on the bottom, showing me what I have.

What I have is a long, paramecium-shaped, black ink-stain on the scan where there should be brain tissue. *It's the effects of a stroke,* he explained. I don't understand, I remember saying; I never had a *stroke.* And then it turned out that this lacuna—or lake, as the neurologists liked to call it—had been visible on a scan for an unrelated workup from many years ago, though the radiologist hadn't noted it then. An old "event," then, probably from around the time I was born, as I'd been in the NICU for a week—but who's to say.

So, relief! Sweet, sweet relief. No tumor, no degenerative neurological disease. I was the same as I'd been before—but not entirely. On one hand, I was grateful that I'd adapted to my vision, which to most would appear flat and fragmented. But with the knowledge of what I wasn't seeing, I felt newly disoriented and vulnerable. My perspective on what I was seeing and what I wasn't kept shifting and refracting, and I started running into things. A friend and I lost our way biking through the steep windy streets of a neighborhood where everyone seemed to have something wrong with them; a middle-aged man with a milky-white cataract in one eye and raw, red-rimmed eyelids stopped to give us directions, stood and watched us pedal away.

I went in to see my internist. I was used to watching him land for a quick minute, perched against the counter, while he joked and scribbled a note in my chart before flying off to his next appointment. This time, though, he sat on a low stool and looked up at me evenly. He didn't betray any impatience as we talked while the tide of patients rose steadily in his waiting room. How could my surgeon have left me worse off after two operations? I asked. I don't know, he replied. But, I asked again, this time more plaintively: if an ophthalmologist is supposed to care for my vision, how could he not have looked to see what I see? I don't know, he repeated. It was probably—he laughed apologetically, took off his glasses and looked away, rubbing his eyes—an oversight. Shouldn't the radiologist have found this lake in my brain before, though? Seems like he should have. How can I trust in medicine now—how can I trust that I'll be cared for? Maybe you can't, yet, he said quietly, as I sat on the examining table, trying to look squarely at uncertainty.

Not many people experience something like this, he tells me—to learn that what you thought was the world is in fact only half of it. And it's true, the oddity of it. Now that I'm aware, I'm stunned at what I don't see. It reminds me of skeptical notions I worked through in childhood: Imagine that there is a flaw in your perceptions or your thoughts, but the flaw masks itself so that you aren't aware of it; isn't it possible that your view of the world could be skewed in such a consistent way that what you perceive and think still appears coherent? It is Descartes's evil demon, deceiving you with an illusory body, an illusory world. And sure enough, now the thought-experiment has come to pass. Just as I assumed that my flat, non-stereopic vision was normal when I was a child, now I learn that I haven't been seeing what I am supposed to see. I have been missing everything to the right of center, all this time. Just imagine.

Then again, seeing what I don't see throws into focus everything that remains. Walking down Broadway one afternoon during these confusing autumn months, I pass a blind man tapping a white cane on the sidewalk, followed by a dwarf. I look up and laugh out loud—I *get* it, I see!—that of all of us walking along—even along this one street—I am so far on the fortunate end of the spectrum. So very, very far on that end.

But then I put my head back down so nobody will see me looking at them crookedly. I glance around covertly and can't help but think: what about all these other people here who have nothing *off* about them, no block between themselves and others? When the conductor spoke to me on the train I looked out the window toward the blur of trees lining the track so he wouldn't see my defective eyes, and I could hear his voice harden in resentment. Then came the sudden flame of his anger—*Why won't you look at me?*—and then my confusion, and finally the sizzling shame when I realized what I'd done. I'm so sorry.

Not to have straight eyes, and to want them, sends my thoughts into the crumpled spiral of a Proustian paragraph weaving in upon itself, forming clause upon sub-clause of metaphor, layering an ever-growing wasps' nest of impacted observations and self-analysis and questions until the hulking, flaking structure breaks off the tree limb and falls under its own weight. This is not a reason not to engage fully with the world—why pay any mind? Would I even want to know anyone who would judge me for this? Isn't it hubris to try to train a bending branch back onto a straight trellis—what if it breaks? After all, isn't this wayward eye part of me? In wanting it fixed, do I betray my singular, imperfect self?

And yet: I *do* want this. I want to look at the world, and I want it to see me. Please. Please, is this something I can have?

I take the train back up north just after New Year's. My eye has been migrating steadily outward toward my ear, leaving a new blind spot in its wake; it needs another re-operation. As I stare at the ceiling tiles in the OR and the anesthesiologist feels my hands for veins, I can sense the brown-eyed surgeon hovering on the edge of my field of view, a tall solicitous figure in blue scrubs, watching and clasping his own hands together in front of him.

He operates and finds that one of the muscles had slipped off the eye after the previous procedures—most likely it hadn't been properly reattached by my hometown surgeon—and it takes several hours just to find it. The recovery is hard, a deep, radiating pain through one side of my head. Later, when I read the operative note, I'm impressed by the graphic narrative arc: searching for the muscle, probing farther and farther back into the orbit, nearly despairing of locating it; finally spying it, encased in scar tissue, scarcely identifiable. A salvage. Several days after the surgery, though, the recovered muscle tightens unexpectedly and pulls my eye far in to my nose, much worse than it's been since I was a baby, and I feel, finally, defeated.

I don't want anyone to see me. I wear sunglasses, even in dark bars, even at night. I'm ashamed to be so ashamed. I take the train back up north, but there's little we can do while the muscle heals. I take off my sunglasses and let my surgeon examine my eye. My stubborn, obdurate eye. We stand talking in the brightly patterned hallway after my appointment, and I watch him literally wringing his hands. I should really stop wringing my hands, he says, looking down at me worriedly.

On the ride home I look out the window and watch the slow snow through my dark glasses. I was wrong to want this, I think. *I brought this upon myself.* Back home, I run along the river in the morning, and I keep imagining that something is striking me from my blind side, and that each time I recover it comes again. At night, I dream that I am dropping through a series of trap doors. I write to my surgeon and describe these visions, though they seem unfairly clichéd (could my unconscious not come up with anything more original than *falling,* I think petulantly.) I understand, he replies—with all that's happened, I thought you were seeing me as Lucy pulling the ball away from Charlie Brown again and again.

During this long winter of waiting we write now and then, and threaded through the clinical terms—*prism diopters, esotropia, fat adherence syndrome*—are metaphors passing back and forth. At first, they are practical: describing a stiff extraocular muscle as a band of tough leather, or a scratch in the eye's cornea, mending in from its edges, like an island surely swallowed by the rising, healing sea. Then the threads begin to tease just a bit, tropes describing tropes. I complain that I've been infected with his Simile Disease, just as butter sitting uncovered in the fridge absorbs other scents. Sorry to contaminate your butter, but those figures of speech do come in handy from time to time, don't you think, he replies. As long as they don't get overextended, because then the butter starts to go rancid and it clogs the intake filters and the refrigerator starts using too much electricity and the city has to have a brownout—yes, it's like that.

There is humor here, and a way to help pass the slow months. But there is also something more, something about self-consciously unfurling these descriptions and exposing them for what they are. Because of *course* metaphors fail. Language fails to describe experience—it fails spectacularly, comically. One thing that is supposed to represent another is never perfectly aligned with it, in the same way that my two eyes, which are supposed to look at you, instead point in different directions, and in the same way that the French term for cross-eyed, *louche*, also describes *something that is not what it appears to be*. It is here, in these failures, that experience reveals itself. It is only when vision breaks apart that one can see what is inside, what the mechanics of sight and communication truly are. Just as it is only when I have fallen backward through these trap doors that I look up to see my lanky metaphor-minded surgeon clambering down to stand next to me. We have slipped through the definitions of language—doctor, patient, defect—and in this clearing beneath are two people, one helping the other to see.

In the spring, finally, I take the train back up north again, and we return to the OR. It takes two more long surgeries, two more long recoveries, and many more days of watching and waiting to be sure the eye will stay in place this time. In between the operations, I find myself here in his office as he tries to loosen this same stubborn muscle. Through the metal instruments and gloved hands circling my eye I hold onto his large brown iris like a buoy, his eye concentrating so closely on mine, nothing else holding me in place except for his gaze. Later I will think, I still don't know what is certain. I don't know what words to use, to thank you. And all I can do, as language fails once again, is this: to tell you what I've seen, looking back.

Elective

ONE

WHEN I WAS sixteen, I had minor elective surgery. My mother wanted me to have it. I went along. Call it cosmetic surgery. Plastic surgery. Aesthetic surgery. Reconstructive surgery. Rhinoplasty. I had a nose job. I had it done. Let me be straight about this from the start. It was not traumatic (or only traumatic in the sense that any surgery is traumatic: there was cutting; there was blood; there was, in the aftermath, an enormous headache). I do not regret it. If I had it to do over again, that is if I were sixteen and had it to do over again, I might do it again. But I am not sixteen. I am sixty, with a turned-up nose.

My doctor's name was Yarmo. Swarthy skin, thick hair; he was tall and magisterial. Googling him now, I come across his obituary, where by last accounts he appears to have done well for himself, an avid member of several yacht clubs. And why not? Plastic surgery is lucrative. In 2011 plastic surgeons made, on average, $270,000, compared with about half that for family practitioners. A report on mdsalaries.blogpsot.com boasts that in California, plastic surgeons could earn upward of 4 million dollars a year: "All depends on how busy you wanna get!"

When I had my surgery, we were far from rich. We lived in a modest ranch house on the North Side of Chicago. My father sold furniture out of model homes. But we were not struggling like we had been either. We no longer had to slink around Community Discount World to buy clothes, hoping we wouldn't see anyone we knew. We were coming up out of the financial doldrums. There was spare cash to invest in my appearance: braces, dermatology, plastic surgery. "Your nose is like an A-frame cottage," Dr. Mark Gorney wrote in a 1976 article on rhinoplasty. We could afford a few improvements to the infrastructure.

There was another doctor, whose name I don't remember, the one everyone in my circle was using then—a circle of middle-class, Jewish, teenage girls—but my mother chose Yarmo because she did not want my nose to look the same as all the others. Each doctor had a signature style, and you could pick the one you wanted. You could shop around. If you weren't going to have your own nose, did you want a Yarmo, for instance, or a Resnick? I took pride in this. My nose job would be different. Or was that a sop I threw to myself (throw to myself even now), something to differentiate me from the myriad other girls from whom I was no different?

Not to say that Dr. Yarmo didn't have an extensive clientele. He did. We flipped through his portfolio. Frontal views and profiles. Befores and afters. It was like looking at pictures of specialty cakes: all those marvelous concoctions. Which one should I pick? (After the surgery, would my pictures be in his book? Would other girls choose to look like me?) The girls we saw all had elective surgery, but there was one photo of a boy who'd been in a car accident. When they brought him in, Dr. Yarmo told us, the boy was a mess. It was a major reconstruction, he said.

Isn't any reconstruction major? I remember when Dr. Yarmo removed the bandages. I wasn't nervous, because I was a child who believed in the promises of adults, but my mother was terrified. What if I looked, not better—prettier, more proportional, more marriageable (that was down the line, of course), more, but not too, Gentile—but worse—misshapen, unaligned, unpretty, unmarketable (again, in the future), still too Jewish? Not to worry. The bandages came off and my nose, though swollen, was pert and sloping upward.

A nose job isn't too bloody, but it's bloody enough. They stuff your nose with packing to absorb the flow. It's not unlike the pad I wore between my legs a couple days after the surgery, when

I got my period. Bleeding out of both ends: that's how, in my misery, I indelicately thought of it. Now it seems oddly fitting, to have had a nose job and my period at the same time, to have simultaneously shed blood in the name of beauty and desire. Sixteen, the age of sexual consent in many states ("Tonight's the night I've waited for / Because you're not a baby anymore," Neil Sedaka sang in "Happy Birthday Sweet Sixteen"); the age Sleeping Beauty, placed under the spell of a wicked fairy, was destined to prick her finger on the spindle of a spinning wheel and die; the age you can get your driver's license in Illinois; the age you can get married with parental consent in Illinois; the age you can donate blood in the United States; the age a girl's nose, 90 percent formed, according to most plastic surgeons, is ready for reshaping.

Susan S., Barbara A., Laurie F. These are the names, decades later, I recall. But there were more of us, a sorority of fifteen-, sixteen-, seventeen-year-olds who, between July 4 and Labor Day, when school was out, slipped into hospitals for surgery. Our brothers had been bar mitzvahed; we, bat mitzvahed or not (many families considered it unnecessary for their daughters to have a religious education), had nose jobs. My friend S., whose CV includes both a nose job and a bat mitzvah, refers to this as the scarification of the nose; along with sweet-sixteen parties, our ritual entry into womanhood. (As for me, I skipped the party and went straight to the operating room.) Some of us claimed a deviated septum, to counter accusations of vanity, but it was a cover no one took seriously; others just seemed to disappear for a few weeks, sans explanation—like girls who fled to homes for unwed mothers—only to resurface miraculously changed. All of us, when the swelling went down and the bruising faded, abandoned the guise of secrecy and proclaimed the news on our persons.

What did it look like before? A few non-Jewish friends want to know. (The Jewish ones don't ask; they've either had a nose job, considered having one, know someone who's had one, or figure it goes without saying why someone, a fellow Jew, would have one.) What they really want to know (those whose curiosity outdistances decorum—not that I mind; after all, I'm the one who's brought it up in the first place, the one, with a frisson of pleasure, who's dangled this titillating tidbit) is what was wrong with it. They want to be able to compare the bad, uncorrected nose to the good, fixed one, inasmuch as anyone can do that without having actually seen the original. Because surely I must have had a defect, a feature that needed correcting. Surely I would not have done such a thing merely for the sake of vanity, or worse, assimilation. (My friend L. can't imagine voluntarily submitting to plastic surgery, especially because she had to have surgery after a dog bit her face.) But I'm not much help in that department. I don't remember what I looked like. I don't have, lodged in memory, an internal picture of myself. Discounting the usual childhood shots, there's only one pre–nose job photo to remind me of who I was. In that photo (fourteen? fifteen?) I look prematurely professorial. Blazer, turtleneck, pageboy. And a nose that lends me a certain snobbish gravitas, a bulbous nose that turns what little I have of a smile into a sneer.

My mother'd wanted to get hers done. She'd wanted to but didn't. She had a beaky nose that trailed down her face and made her look melancholy. L., who'd seen a picture of her in my study, said she'd been a real looker. A looker. I hadn't thought of her that way. It made me, with a start, reconsider her—yes, my mother, with her high forehead and serious eyes and aquiline nose, had been beautiful. "The Lush Kid," her nickname in her high school yearbook. Winsome, my father had once called her. I suspect she didn't get her nose done because she couldn't afford it. She would have had to pay for it herself: her mother was a homemaker; her father an immigrant sign painter. And how much money did she make near the end of World War II, when she was twenty years old and working for a downtown dress manufacturer? Enough, perhaps, to buy smart clothes, but not enough for a nose job.

But maybe there were other reasons as well. In 1944, during May and June alone, 476,000 Hungarian Jews were sent to Auschwitz. On August 23, my mother's twentieth birthday, the headline on the front page of the *Chicago Daily Tribune* read, YANKS GAIN 65 MI., REDS 38 / REPORT ALLIES AT BORDEAUX. Paris was liberated two days later. Perhaps, as America fought the Good War, as the war dead—Allied and Axis, military and civilian, Jew, homosexual, gypsy, Communist, Socialist, trade unionist, Free Mason, Jehovah's Witness, deaf, blind, insane, depressed, schizophrenic, epileptic—neared 70 million, it would have been indecent for Harriet Alter, a young American Jew on the West Side of Chicago, only child of Goldie and Leo Alter, to have a nose job. Or maybe not. If German Jews were having nose jobs in the years leading up to the war—and they were, for a complex mélange of reasons including vanity, mental health, and survival itself—perhaps it was no more unseemly for one of their counterparts across the ocean to consider the same. In the end, however, this is all just the speculation of the next generation, of a baby boomer who, over twenty years later, along with a significant number of her peers, was drugged and numbed and went under the knife.

TWO

My nose job has made me a snob about nose jobs. Not long ago I saw a woman in a department store who'd had one, and in a triumph of recognition, I hurried to tell Ann. I felt as though I'd outed her, a smart, savvy-looking woman with sleek dark hair and a rhinoplasty. I know your secret, I thought. I can tell. I can, in fact, spot many of them, something about the sculpted contour of the nostrils, the precise, carved look, or maybe that's mostly the old-time nose jobs, the ones from my generation. (But this woman was not my generation; she was much younger.) Some noses just don't fit. Barbara A.'s was too short. Susan S.'s, my cousin's, too snub. Mine's snub too. Maybe snub was in then, snub the standard, but it didn't suit our faces. (While snub was the fashion in 1967, snub, circa 1887, when Dr. John Orlando Roe, in Rochester, New York, introduced modern rhinoplastic techniques on pug-nosed Irish immigrants, was considered by some to be a deformity. It indicated, according to Roe, weakness of character, a view no doubt influenced by the widespread anti-Irish sentiment of the nineteenth century.) Or is the face like a palimpsest, the old face constantly showing through the new one, the old nose poking through, so I can't totally erase the picture of what was and replace it with what is? Is my old nose a shadow nose, still lurking, still big and bulbous, asserting its Semitic rights over its pert replacement?

In a horrible twist on this notion, such reclaiming is, in fact, what happens in German writer Oskar Panizza's tale *The Operated Jew* (1893). Jewish medical student Itzig Faitel Stern is an alarming specimen of human being. His chin is drilled to his chest, his legs are bandied, his speech is a mixture of "fatty guttural noise [and] soft bawling." He waddles when he walks and gestures wildly and effeminately when talking. His mouth foams with saliva. How can he live this way? How can he, a man "who emanated directly from the stingy, indiscriminate, stifling, dirty-diapered, griping and grimacing bagatelle of his family upbringing," make his way through the chaste and noble corridors of Heidelberg society? He can't. And so he submits to a series of grueling procedures: multiple surgeries to straighten his bones, speech therapy, a four-week drug regimen to lend his skin a Teutonic tint, hair coloring and straightening, and lastly, to complete the job, eight liters of Christian blood (a reference to the infamous charge of blood libel). Voila! Meet Siegfried Freudenstern, who, thus transformed (apparently no one realizes this is the former Faitel Stern), is ready to take a German wife and propagate the race. Only, something goes awry at the wedding. Siegfried has too much to drink (Jews, of course, can't hold their liquor), and under the enormous pressures of disguise and conformity, and with no small amount of lip-smacking vindictiveness, his cankerous Jewish self explodes. The changes, never really permanent in the first place, come undone. His speech takes on its former patterns ("I vant you shood know dat I'm a human bing jost as good for sumtink as any ov you!"); his posture assumes its "crippled looking compulsive formation"; his hair curls and turns black; and he collapses in a pool of vomit. Itzig Faitel Stern. Operated, manipulated, mocked, destroyed, he is once and for always The Jew.

THREE

Until recently, I thought Jimmy Durante was Jewish. Durante. The son of an immigrant Italian barber. Christened James Francis Durante. Grew up on New York's Lower East Side. An altar boy at Saint Malachy's. I thought he was Jewish because of his nose. I was disappointed to find out that he wasn't. I'd wanted him to live up to my expectation of a big-nosed Jew, a lovable wise guy I could claim as one of my own; and if I couldn't claim him as one of my own, then at least I'd point out that he grew up among the Jewish garment workers of the Lower East Side, as if, ridiculously, their Jewishness and their nozzles could somehow have rubbed off on him.

What does this say about me? That I've succumbed to certain assumptions about My People? That I, a former big-nosed Jew, am full of tired preconceptions? Outsized pride? Am I claiming the same kind of in-house privilege that leads Blacks, when speaking of their own, to say *nigger*, or queers *queer*?

Is the iconic nose iconic because it's Jewish?

Durante, Catholic and not one of my people, nicknamed himself Schnozzola. Schnozzola/Schnozzle/Schnoz. Probably an alteration of the Yiddish *snoyts,* "snout, muzzle," from the German *Schnauze* (the German schnauzer, a vigilant, sober dog, very loyal to its handler, not led astray by bribes). And/or the Yiddish *shnabl,* "beak." Durante reportedly said that after his schoolmates mercilessly taunted him, he'd "go home and cry. I made up my mind never to hurt anybody else, no matter what. I never made jokes about anybody's big ears, crossed eyes, or their stuttering." Instead, he made jokes about himself. "My nose isn't big," he said. "I just happen to have a very small head."

Milton Berle, on the other hand, *was* Jewish (born Milton Berlinger, son of Sarah and Moses Berlinger), with a noteworthy nose, or at least a nose he deemed noteworthy enough to have bobbed, as he put it (*bob:* to cut short; dock; often with *off:* as, to bob or bob off a horse's tail). One report says he broke his nose during a comedy routine, but Berle, in his autobiography, makes no mention of this. All he says is that the tip turned down when he smiled and he decided to fix it. Pleased with his new look, he gave nose jobs as presents to friends, one of whom, in a strange bit of only-in-America nomenclature, dubbed him Santa Schnozo in return. "I cut off my nose to spite my race," Berle joked during his routine.

One Rosh Hashanah I went to Manny's diner for breakfast and ordered pigs in a blanket. I'm a hand-wringing kind of Jew: I was taking the day off from work for the Jewish holiday and eating sausages wrapped in pancakes and feeling uneasy about the whole arrangement. About my truancy, the pork, my observance/nonobservance of the holiday. I had no plans to go to temple. That word is a holdover from my childhood, when we called this place we rarely went to *temple,* and a holdover from the Temple that used to stand on the Temple Mount in Jerusalem. I took the day off from work because I didn't feel right about working on the holiday (although I didn't feel right about not working, either) and because my boss, owner of the print shop where I ran the offset press, was also Jewish.

Eating the sausages, I experienced a queasy pleasure. They were forbidden (*trayf,* impure, not Kosher), but they were greasy and good. They felt more forbidden because it was the holiday and I was making a feeble attempt to observe it. Forbidden? I'd always eaten forbidden food. In our house it was not forbidden. In restaurants it was savored. Shrimp scampi, ribs, egg rolls, BLTs. Although we'd always shied away from ham and pork chops. That seemed to cross some line. There was *trayf* but good, and then there was too *trayf*. Ham and pork chops were too *trayf*. On this day, this Rosh Hashanah, I slathered the pancakes and sausages with syrup. The sweet and the savory together. Possibly I read the newspaper as I ate. Then someone I knew came into the diner. Someone about whom, all these years later, I can hardly conjure up a single detail. But in that instant I felt caught. Busted. A true truant. A Jew on the lam. She must have said something about my not being at work that day, and I must have said something about the Jewish holiday. Because then she said, *I didn't know you were Jewish.* And I thought, must have thought, think now, *What did she mean by that?* Certainly there was an awkward silence. *Well, I am,* I said, said self-deprecatingly and defiantly at the same time. The self-deprecation and the defiance seemed to cover all the bases. Shame, embarrassment, confusion, defensiveness, anger, pride. Perhaps I added an offhand laugh. Can't you tell? I thought. What's it to you? And now that you know?

Am I being too harsh? Ungenerous? Did I misunderstand her? Valid questions, but another tugs at me more urgently. Why didn't she know? Why didn't she recognize me? I want people to know I'm Jewish. I want to be taken for who I am. I bear the signs of assimilation: an indeterminate surname— when my father and his brother changed their name from Shinitzky to Shinner, the change, in and of itself, could have been read as a possible sign of Jewishness. In 1922, as part of an effort to lower the percentage of Jewish students, Harvard revised its application form. One of the new questions was "What change, if any, has been made since birth in your own name or that of your father? (Explain fully)"—an improbably Irish-sounding first name, an omnivorous diet, a nonreligious lifestyle, a snub nose. I didn't get a nose job to look less Jewish. (But isn't that what everyone says, patients and practitioners alike? "I wanted to look prettier and my nose was a sight in any language, but I wasn't trying to hide my origin," singer and actor Fanny Brice once protested. Or, as Drs. Dennis P. Cirillo and Mark Rubenstein put it in *The Complete Book of*

Cosmetic Facial Surgery, "If your daughter does wish to belittle her background, other signs of this intention would, no doubt, have appeared by now.") I didn't get a nose job for any particular reason I could have articulated at age sixteen, except, like Brice, to look better. I got a nose job mostly because it was expected of me. Expected of me, particularly, and expected of someone like me: young, female, Jewish, with fair-to-middling, meaning subject-to-improvement, looks, in the middle of the twentieth century. Small wonder, perhaps, that my acquaintance at Manny's hadn't tagged my identity. Had the evidence been erased from my face, surgically removed? Would my big and bulbous nose have delivered the news?

The Jew's nose, my nose (my former nose, or some rendition thereof), has long existed in the popular imagination. Over the centuries it's been described as humped, hooked, hawkish, club shaped, crooked, convex, and in the children's book *Der Giftpilz* (*The Poison Mushroom*), published by Nazi propagandist Julius Streicher, "like the number six." "It indicates considerable Shrewdness in worldly matters; a deep insight into character, and facility of turning that insight to profitable account," wrote George Jabet, aka Eden Warwick, in *Notes on Noses,* a nineteenth-century physiognomy text. This nose, real, satirized, or an anti-Semitic construction, has circulated as common currency. In an unintentionally comic attempt to refute it, JewishEncyclopedia.com cites a study which found that among modern Eastern European Jews, the majority of noses are not arched or hooked but "straight, or what is popularly known as Greek."

Some version of the Jew's nose exists in my mind as well. My friend M., who herself had a nose job and is resentful about it, or resentful of the pressures that forced her into it, not the least of which was an insistent mother, recently went to Israel to visit relatives. I have long admired M.'s nose, which is straight, strong, and subtly authoritative, and was surprised to find out she'd had it done. I was surprised because it looks, post–nose job, Jewish. In talking about her trip to Israel, about family relations, the West Bank wall, the occupation, amid all this parsing of the complexities of our time, M. stopped for a second and laughed and said, in a voice suggesting she might admit this only in the company of friends, "It was a pleasure to see all those distinctive noses."

FOUR

Did M.'s nose job do the job? Did mine? What do we, who find ourselves allied in the strange confederacy of nose-fixers, aspire to? A nose that conforms to a facial angle of thirty-three degrees and is the approximate angle of Venus de Milo's nose, which was the feminine ideal suggested by fin-de-siècle German Jewish surgeon Jacques Joseph, a pioneer of modern plastic surgery? (For a man, the preferred angle is slightly bigger, implying that a man's face can, and indeed should, carry a more prominent nose.)

What about passing? Dr. Joseph's first rhinoplasty patient, a man of unknown origin whose large nose had led to public ridicule and subsequent depression, indicated that after the surgery his distress subsided and he was able to move around unnoticed. Sixteen-year-old Adolphine Schwarz, another of Joseph's patients, had her nose done in 1934, a year after the Nazis came into power. Her older brother, "suffer [ing] from a Jewish nose," had already undergone the operation. "I . . . had a nose that bothered me, especially after the Hitler regime started," Schwarz noted. "So it was decided that I too should have my nose reshaped." Schwarz's nose job, of course, did not protect her. Four years later, she and her family were forced to flee Munich and their famous restaurant, whose patrons had included Albert Einstein and Charlie Chaplin. Her parents died in Auschwitz, while Schwarz and her husband eventually resettled in the United States.

For some, such as Milton Berle, the stakes were much less

high. He hoped a nose job would help boost his career. "I liked the new look so much," he wrote in his autobiography, "I decided it would make all the difference when—and if—the next shot at the movies came." Indeed, the next shot came shortly after, in 1941, with *Tall, Dark, and Handsome,* which won an Academy Award for Best Original Screenplay.

Is a nose job successful if the person didn't want it in the first place? If she's indifferent about how it looks but unhappy with what it means? M. doesn't say whether she is happy with her nose, but she is clearly troubled by her nose job. Troubled, I suspect, by the prevailing social attitudes that judged her original nose unattractive, unfeminine, and, of course, demonstrably Jewish. Troubled as well by her understandable inability, as a teenager, to stand up to those judgments. For my part, if I'm troubled neither by my nose job nor my nose, I'm uneasy with my passive acceptance of the anonymity it has given me. Anonymity lets others fill in the blanks. *I didn't know you were Jewish,* my acquaintance at Manny's diner had said. *You don't look Jewish,* she might as well have said. On what was she basing her assessment? Sure, I want to be taken for who I am, but like Adolphine Schwartz and all the others who "suffered from a Jewish nose," I don't want to suffer too much for it.

What do the rabbis have to say about nose jobs? I have never been in the habit of consulting rabbis, except in times of death, when I've looked to them for last farewells—even if, in the manner of Durante, they've had not much more to say than "Goodnight, Mrs. Calabash, wherever you are." But, as is also my habit, I seem to be interested in their prescripts after the fact. I guess I want to know where I'd fit in, if I'd fit in at all, where my actions fall in their grand accounting.

As it turns out, the rabbis approve. They give a nod to my nose job. The Jews are a practical people. There are laws that govern all aspects of Jewish life ("one must not walk 'four cubits' upon rising in the morning until after having poured water three times . . . upon the fingers of each hand alternately [so as to rid the body completely of the evil spirits that entered the body upon going to sleep at night]"), and then there are the allowances. Four laws, in particular, are relevant to nose jobs: those that prohibit indulging in vanity, wounding the body, endangering one's life, and tampering with God's handiwork. (In an odd and macabre twist of this idea, neo-Nazi Jonathan Preston Haynes murdered plastic surgeon Martin Sullivan in 1993 because Sullivan, by operating on patients and altering their God-given, presumably Semitic features, "dilute[ed] . . . Aryan beauty.") But these mandates may be set aside, according to various twentieth-century Jewish legal scholars, if a woman's nose—her deformity, as Rabbi Immanuel Jakobovits puts it— "make[s] it difficult for [her] to find a matrimonial partner or to maintain a happy relationship with her husband." Relief of emotional pain overrides the ban on wounding the body, adds Rabbi Yaakov Breisch. The anguish of not being able to find, or keep, a suitable partner, Rabbis Menashe Klein and Yitzchak Yaakov Weiss concur, is sufficient reason to allow cosmetic surgery. If surgery helps a girl find a mate, it is self-healing, not self-wounding, says Rabbi Moshe Feinstein.

In other words, make an appointment.

As for men, the rationale is different but the outcome the same. Deuteronomy 22:5 reads: "A woman shall not wear a man's garment, neither shall a man put on a woman's garment; it is an abomination before God." Men cannot dress like women, masquerade as women, take up female behaviors, be one of the girls. Therefore, in a world where cosmetic surgery is typically associated with women and vanity thought to be an especially female and corrupting trait, the rabbis might be expected to rule against nose jobs for men. But more pressing social factors are at play. What about a man's role as provider? (He was a good provider, my great-aunt used to say about her philandering husband.) A man may have a nose job, Rabbi Jakobovits argues, if his deformity prevents him "from playing a constructive role in society and . . . maintaining himself and his family in decent comfort." In this view, a nose job may be the ticket to a middle-class life, a life of decent comfort; and although that seems laughable at first, or it did to me, research shows the rabbi might be right. According to a 2005 study, below-average-looking men suffer a "plainness penalty" in the labor market: they earn 9 percent less than their better-looking counterparts. Good-looking men, says the study, get more job offers, higher starting salaries, and bigger raises. Looks pay.

FIVE

There is another picture, my post–nose job picture, my after shot. It was taken at a sweet sixteen. In my tight-fitting dress and new nose, I'm practically bursting out of the picture. It is 1967, ten years before the premiere of *Roots,* five years before *Ms.* magazine debuted, three years before Kent State, two years before I went to college, one year before Barbra Streisand—perhaps the most famous Jew who never had a nose job, whose nose was described in *Ladies Home Companion* as "overly-prominent," in *Time* as a "shrine," in *Newsweek* as "absurd," in *Life* like "a witch['s]," in the *Saturday Evening Post* like "an eagle['s]," and in *Life* again as "like Everest, There"—starred in *Funny Girl,* a film about Fanny Brice. In the photo, my skin is shiny, my nose is swollen, my smile is slightly pleading. My dress is the color of well-fertilized grass. Next to me is Barbara A., the daughter of Holocaust survivors who tried to flee Poland on the back of a horse cart but were caught and sent to a camp. She hasn't had her nose done yet, hasn't had the hump on her bridge, inherited from her father, chiseled down. But she will, by the time *Funny Girl* opens and Streisand, as Best Actress, wins an Academy Award.

STEVEN CHURCH

Speaking of Ears and Savagery

ROUND I

Like Dempsey, he has the power to galvanize crowds as if awakening in them the instinct not merely for raw aggression and the mysterious will to do hurt that resides, for better or worse, in the human soul, but for suggesting the incontestable justice of such an instinct. . . .
—Joyce Carol Oates, "On Mike Tyson"

On June 28, 1997, in Las Vegas, during the rematch fight between Evander Holyfield and Mike Tyson—a fight billed as "The Sound and The Fury"—things weren't going well for The Fury.

Tyson had already been beaten badly by Holyfield in the previous fight, suffering a TKO, or "technical knockout," in the eleventh round after a sustained pummeling. That match had shown Tyson to be vulnerable, and he looked every part of the sports cliché "a shadow of his former self." He made excuses afterward, claiming Holyfield had used intentional head-butts to cut and daze him when the two fighters entered into a clinch. Most people believed Tyson had lost his edge, had grown fat on the largess of his life or been corrupted by the influence of Don King—all of which was true.

Tyson's complaints, however, were not totally without merit. Holyfield had long been known as a master of the subtle head-butt, a tactic that, while common, is hard to spot and is potentially devastating. A head-butt in boxing is not the exaggerated forward strike you see in professional wrestling or movies, but a more subtle tactic of close-in, hand-to-hand combat, a swift

58

strike with the crown of the head to the thin-skinned brow, cheek, chin or forehead of an opponent. Done correctly, discretely, it can quickly disable an opponent by knocking him stupid or by causing swelling to the eye or excessive bleeding, which blinds the fighter.

Intentional head-butting is against the rules of boxing, a violation similar to a punch below the belt. The head-butt is considered a "dirty" tactic—just the sort of trick you'd think a man like Mike Tyson, not a man like Evander Holyfield, would use to gain an advantage over his opponent. Perhaps because of this, all of Holyfield's head-butts were judged to be "accidental."

As the rematch fight entered the second round and Tyson's furious efforts to slow the fundamentally sound and patient Holyfield seemed fruitless—several clean punches failed to deter Holyfield's steady advance—the two men entered into a clinch, and another head-butt from Holyfield opened a sizable gash above Tyson's right eye. With blood streaming down his face, Tyson complained bitterly to the referee and admitted later that he was dazed and scared, feeling vulnerable, but the referee, Mills Lane, ruled the head-butt was unintentional.

Angered by the head-butt and Lane's refusal to intervene, Tyson came out for the third round and unleashed a barrage of punches at Holyfield, but his rally barely fazed the champ. The two men clinched up, and again Holyfield head-butted Tyson, who, at this point, became convinced Lane wouldn't protect him. He was desperate, angry, and determined to defend himself.

Tyson, despite all his fury and bluster, always spoke with a lisp, and his characteristic high-pitched, nasally voice made him sound like a man-child, a curious mix of innocence and aggression. Tyson always, always fought as if he'd been beaten back into a corner and told to stay there. Like everyone else in the world, I'd come to love Tyson for his naive ferocity, for the brutality with which he dispensed opponents, often exploding as if he'd been unchained and turned loose from his corner. He didn't just beat his opponents; he went out there in his black trunks, black shoes, and short socks, and he humiliated his opponents. We loved every terrifying second of the carnage, especially when he threw his uppercut, a punch that seemed capable of knocking a man's head clean off. Most of us wanted him to destroy Holyfield, to show the world that Iron Mike was still a force to be feared. But something was wrong from the beginning. Tyson wasn't himself, wasn't the Fury we expected. Instead, he became something else entirely, something much worse—a mirror, a reflection of our own bloodlust, a vessel for our collective savagery. Tyson lost the artistry that made his brutality beautiful. It disappeared into the fog of fear.

Lane separated the two men, and the fighters exchanged a few punches before locking up again. As they did, Tyson spit out his mouthpiece. When Holyfield's head came up, Tyson twisted his neck, tucking into the side of his opponent's face almost as if to kiss him on the cheek or nuzzle his neck.

Tyson opened his mouth wide. He bit down hard on Holyfield's upper right ear, severing the helix, the outer section of cartilage. Holyfield jumped back, shoving Tyson, who spit out the chunk of ear onto the canvas. As Lane tried to figure out what had happened, Holyfield hopped around the ring, gesturing at his head with his glove as blood poured from the wound.

The boxers were sent to their corners. Everyone watched. Then, perhaps caught up in the moment and not fully aware of the extent of the damage to Holyfield's ear, or perhaps so sucked into the adrenaline of blood-sport that he was blinded to the reality of what was happening, Mills Lane allowed the fight to continue. He wanted it to continue.

After exchanging a few punches, the two fighters locked up again. This time, Tyson bit down on Holyfield's left ear—not as hard as the first bite but still hard enough to cut and leave a mark and send Holyfield jumping back, flailing his arms hysterically and pointing first at Tyson and then at his own head with his cartoonish red gloves.

Lane again sent the fighters to their corners and finally ended the fight, waving his arms in the air and disqualifying Tyson, who exploded in rage and rushed at the Holyfield corner, throwing punches at anyone who got in his way. Soon, the ring was flooded with thick-necked sheriff's deputies in beige uniforms. As Tyson exited the ring, boos erupted from the crowd. The whole place seemed to surge and pulse with adrenaline. Tyson bulled his way toward the exit, and a fan threw a water bottle at him. He jumped over the barrier, charging into the crowd, screaming profanities and pointing at people, raging at anyone near him. Members of his entourage and security personnel dragged Tyson out of the stands and pushed him toward the exit.

Unable to look away from the television screen, I watched the spectacle unfold from a safe distance. Red gloves. Black men. The bloodied helix. The noisy physics of Tyson's fury. And part of me wanted to be waiting in the locker room as Tyson came back. Part of me wished I could get close enough to touch his unchained violence, but most of all, I wanted to see the severed ear, the bloodied helix, lying on the canvas or cupped in a white towel, nested in a bucket of ice.

ROUND 2

I just want to conquer people and their souls.

—Mike Tyson

David Lynch, the mind behind such cinematic creations as *Twin Peaks, Eraserhead, The Elephant Man, Dune,* and *The Lost Highway,* also wrote and directed 1986's *Blue Velvet,* a movie that changed forever the way I thought about ears.

In particular, it changed the way I thought about ear-related savagery, about the meaning of severed ears. Such thoughts aren't something you always want. Perhaps you unwittingly stumble into Lynch's vision. You're simply watching a young man walk through a field of overgrown grass in the opening scene of *Blue Velvet,* and you have no idea what to expect. Perhaps you watch him closely, this tall, pasty-faced man in Lumberton. Everytown, America. Innocent but curious. A man, but just barely. Jeffrey Beaumont. He's cutting through this overgrown lot after visiting his dying father in the hospital. He's not expecting any complication to his life. He has no idea how quickly things can change, how one small discovery in a field of secrets can crack open his world.

Jeffrey finds an ear in the grass. A human ear. Severed from the skull, it becomes like a window, or a rabbit's hole. Lynch said of this scene, "I don't know why it had to be an ear. Except it needed to be an opening of a part of the body, a hole into something else. . . . The ear sits on the head and goes right into the mind, so it felt perfect."

The brilliance of the scene is captured in the dilemma it hands off to the audience: The question put before them is "What would you do?" It implicates us in everything that follows—the twisted search for answers that leads to Isabella Rossellini and the infamous scissors, Frank and the mask, hiding in the closet, the car ride, the dancing, the drinking, the kidnapped boy, Dean Stockwell, nitrous oxide and Roy Orbison—all of it spinning madly out of control. And it all begins with curiosity, attention to an ear and a lingering question: What would you do? Could you stop yourself from falling, too? It's strange how a subject overtakes you. This thing that Jeffrey cannot leave alone, this thread he cannot stop tugging, makes everything come undone. Lynch's camera takes you down into the ear, and you fall into the darkness. Your life is never the same.

Again and again, I fall into the severed ear in this scene, disappearing into its rabbit's hole. This ear replaced Van Gogh for me—the easy, comforting allusion of aberrant emotion Lynched forever after, corrupted and complicated in ways that are so much more unsettling than the anecdotal violence of Van Gogh's severed ear. Now, when I hear "severed ear," it's not golden sunflowers and unrequited love I think of, or even a drunken quarrel between rival artists, but, instead, the mess of Mike Tyson or David Lynch's particular brand of savagery.

ROUND 3

Everyone has a plan until they get punched in the face.

—Mike Tyson

The word "earmark" appears to have originated in the fifteenth century, referring to a notch cut into the ear of cattle or sheep to indicate ownership. When I was eighteen, done with team sports, I pierced my left ear to declare some small measure of independence from my parents' shepherding influence. I marked the other a year later because I appreciated symmetry. I kept my ears adorned for over a decade and only took the jewelry out when I had my first job interview for a teaching position. I don't wear jewelry in them any longer, but I don't need the decoration for them to get noticed. I have large ears with prodigious lobes, which dangle down like tiny saddlebags.

Next time you're in a classroom or a restaurant or some public place, look around and see how many ears you can spot. Pay attention to how people wear them, cover them and show them off, however subtly or unintentionally. Pay attention to size and angle of articulation. Pay attention to gender. Ask yourself—if you can stand the self-consciousness—how much effort you put into your own ear display.

Aside from some new mysterious hair sprouting from them (and inside them!), I don't have to worry much about my ears. I live in a climate where frostbite is a concern only for fruit. I might suffer the occasional sunburn if I'm not careful. Most of the time, though, I take my ears for granted. If they weren't attached to my head, I'd probably leave them in coffee shops and bars, drop them on morning walks, lose them in the cushions of the couch. I'd probably find the dog in the backyard, chewing on one of my ears, or have to scold my toddler daughter for depositing my ear in the fish tank. I'd buy extra pairs of ears—the cheap kind you get at convenience stores—just to have some backup pairs.

I have large saucer-like auricles, maybe a little too big for my head, a little like costume ears. Some people in my family—but not me—have ears that stick straight out from their heads; my dad jokes that they "look like they're driving down the street with their back doors open."

Some people crack their knuckles or chew their fingernails (which I also do when I'm forced to sit still for more than a few minutes). My whole life, I've played with my ears. I can't help it. I like the stiff leathery feel of them; the soft, squishy, peach-fuzzed lobes; even the cheesy stink you get sometimes behind the ear. They feel cool and stiff sometimes, like plastic props or like, I imagine, a dead body; other times, they're warm and pliable as bat wings. If I'm having a really hard time—say I'm at a particularly odious meeting or poetry reading—I'll tug hard on my ears, yanking them down and twisting them, sometimes rocking back and forth in my seat. I'm sure I look like some kind of mental case, but it's the only way I've learned to stay in my seat.

Most of all, though, I like to fold one of my auricles (usually the right one) down and stuff it into the ear canal. If it's cold, the ear will stay folded and stuffed for a few seconds or until I flex a muscle in my face and the ear pops out. I do this sometimes absentmindedly or, if I'm in a class, when trying to sit still and pay attention.

I can also wiggle my ears, causing them to wave back and forth subtly—a trick that unnerves some people and makes others say, "Oh, that's easy," as they begin the pained effort of attempting to do the same, which usually just results in them raising their eyebrows up and down in an exaggerated motion, looking even sillier than I do.

It's not easy. I trained myself to wiggle my ears, practicing in front of the bathroom mirror, staring at myself, pinching and flexing face muscles until I got it right, until I could move them on command. My father can do it, too. As a boy, I stood awestruck before him, demanding he repeat it over and over again, just as my daughter does to me now about my ear-stuffing trick, saying, "Daddy, I want an ear trick," and giggling hysterically when my ear pops out.

The truth is all human ears are somewhat interchangeable. They share physical characteristics, the same language of structure—the "helix" and "antihelix" ridges, the subtle "antihelical" fold between them, a small canyon of flesh; the shadowy "scapha," "fossa," "concha," and the somewhat superfluous "lobule"; and before you reach the "external auditory meatus," you have to negotiate the "tragus" and "antitragus." The differences between one person's outer ears and another's are subtle, slight and usually a matter of inheritance, aesthetics, and centimeters.

We may take it for granted, but the human ear—the first sensory organ to develop in the womb—is responsible for a surprisingly complex mission of protection and service. On one level, the outer ear, also known as the "pinna" or "auricle" (yes, sounds like "oracle"), is basically just a large sound-wave deflector, which keeps noise from zipping past our listening holes, but the ear's shape and design is far from arbitrary.

The curves and folds in your pinna are designed to funnel sound waves down into the inner ear's more subtle and delicate machinery, which lies just beyond the "tympanic membrane." It's here that the more complicated work of hearing is done, where the waves are translated into electronic impulses, which fire neurons and other brain receptors, generating what our brains recognize as sound.

While the inner ear is responsible for our sense of balance and equilibrium (something else we often take for granted until we lose it), the outer ear also works to help regulate body temperature, with our auricles serving as thermometers and heat transference devices. Perhaps most striking, the outer ear is surprisingly, often embarrassingly, responsive to emotion.

If I get embarrassed or nervous or scared, my auricles flush red and feel hot to the touch, erupting from the side of my head like flares or flags signaling emotions I want to keep bottled up. I've often been told I have a terrible poker face. This is true. My ears are part of the problem.

Still, I appreciate their personality. I need my ears, and for more than just their necessary physical functions, more than hearing. I need them for the physical act of thinking, for listening to myself and essaying one tangent or another. Just as my kids did when they rode on my shoulders, I often use my ears as handles, places to put my nervous fingers. I tug and twist them. They are touchstones for me. Talismans. Tangible things I can use to keep my hands and brain busy, to help me find thoughts lurking at the edge of clarity, to keep me grounded and listening for more.

ROUND 4

> *To come to a scene and you see a fellow human being ripped apart, I feel for that.*
>
> —Officer Frank Chiafari, Stamford Police Department

Travis, a two-hundred-pound, fifteen-year-old chimpanzee, lived in a private home in Connecticut for most of his life. His owners, the Herolds, operated a tow truck business, and Travis used to ride along in the truck to help stranded drivers. He was something of a local celebrity, an animal that acted the role of family member and friend. He drank wine, ate steak, and enjoyed many of the finer things of human life, but it seems Travis was also a sad chimp—perhaps even a depressed, anxious, and fed-up chimp, frustrated with his own brand of captivity, tired of the expectations that he be so tame, so unnatural.

On the last day of his life, Travis was particularly agitated and was given Xanax for his rage. It didn't help. He'd gotten outside the house, out of control, and he was angry. His owner, Sandra Herold, called her friend Charla Nash as a last resort. Travis knew Charla and seemed to trust her. She had a calming way. Charla pulled up to the house and stepped out of her car. All she wanted was to get Travis back into the house, to get him to calm down and feel safe again.

But Travis had crossed over. He'd escaped the role assigned to him.

Charla barely made it a few steps before Travis viciously attacked her in the driveway. By the time he'd finished, he'd bitten or ripped off Charla's nose and lips, her eyelids, part of her scalp and most of her fingers. One of the first responders on the scene said he couldn't believe an animal had done that and said it looked as if her hands "went through a meat grinder." For his part, Travis had suffered a stab wound to his back when Sandra Herold plunged a steak knife into him in a futile attempt to stop the attack.

Officer Frank Chiafari was among the first responders. He'd known Travis and liked him. But as Chiafari pulled into the driveway, Travis was no longer endearing, no longer an innocent pet; he'd become something else entirely, a manifestation of savagery unchained, the worst side of nature.

Charla lay semi-conscious and horrifically maimed on the ground. Before Chiafari could react to her condition, Travis approached his car, knocked off the side mirror as if "it was butter," grabbed the handle and yanked open the door. In some awful approximation of a horror-movie scene, Travis stood there covered in Charla's blood, opened his mouth to shriek and bared his teeth.

Chiafari, cowering in his car, shot Travis four times with his service revolver. Travis stumbled away from the car, back into the house, crawled into his bed, and died where he slept most nights.

Afterward, the officer was haunted by what he'd seen and by what he'd done. He couldn't visit a mall or an amusement park without being haunted by images of faceless women. He avoided zoos or the circus or any place where he might see a chimpanzee, an animal that Chiafari knew was capable of perhaps the most human and natural instinct of them all, extreme violence and savagery.

For some reason, Travis didn't touch Charla's ears.

During her appearance on *The Oprah Winfrey Show* in 2009, Charla's ears were, in fact, the only feature that allowed you to recognize the thing on top of Charla's shoulders as human, the only recognizable facial feature on a head that looked more like an abstract sculpture of a head. Charla's eyes had been removed, and she drank through a hole in her face. She had only one thumb remaining.

Charla Nash could still hear just fine. And this, it would seem, was both a blessing and a curse. She still felt like the same person inside, and because she couldn't see or touch her face with anything besides her one remaining thumb, her main understanding of how she looked was gained by listening to other people's reactions. She could only hear what others saw—or didn't see—in her face, and she wore a veil to protect us. She said she still felt like the same person—still felt like a person.

"I just look different," Charla said.

ROUND 5

> My style is impetuous. My defense is impregnable, and I'm just ferocious. I want your heart. I want to eat his children. Praise be to Allah!
>
> —Mike Tyson

Fifteen years later, Mike Tyson is only mildly repentant for biting off Evander Holyfield's ear. Basically, he's sorry he was forced to do what he did. A recent documentary on Tyson's life seems to support his claims that Holyfield used head-butts for a tactical advantage in each of their fights and that, for whatever reason, the referee in each fight mostly ignored the tactic.

Is it possible that Tyson became a victim of his own image—the mad-dog fighter, the Fury? If everyone else believed it, why wouldn't the referees also buy into the hype, the belief that Tyson represented something animal and primal while Holyfield symbolized the Sound, the humble, soft-spoken gentleman boxer who would never intentionally head-butt an opponent?

You can watch videos of the incident online. As bizarre and grotesque as it is to see the ragged tear on Holyfield's ear and the blood pouring down his head, to watch Tyson spit out the chunk of helix, and to think about what it would take to bite through flesh and cartilage, to sink your teeth into someone's ear, it is also strangely unsurprising. Normal. Even predictable and, in some ways, entirely justified if you think about it. You might have done the same if you were in Tyson's shoes. Really, what would you do? How would you defend yourself from head-butts?

Aside from the impulse to self-defense, the instinct to bite is ugly, but it makes sense. It's natural. As someone who obsessively chews his fingernails and his pens, who has watched his babies' faces contort with teeth-pain relieved only by chewing or biting down into something, who has seen a frustrated toddler bite because she can't do anything else, I can recognize that the urge to bite is not always an urge to hurt but sometimes an effort to find comfort and security, intimacy and escape from pain. It's a desperate effort at self-soothing. My daughter has bitten me when we are playing happily, rolling around on the floor. She's not trying to hurt me. She just gets overly excited. She's trying to hold me, to be close to me. She wants a connection closer than touch, wants to feel safe and secure.

Tyson's no child, and I don't mean to infantilize him, but I believe his actions were not so much those of a violent man but instead the existential jaw-clenching of a lonely, frightened human being—desperate, tragic for sure, but not necessarily chaotic or aberrant, certainly not inhuman. To me, watching the videos of the incident now, Tyson looks like a man who wanted to be held. I hope that if I had been there in the locker room, if I'd been a different person, someone in Tyson's inner circle, I would've recognized that need in him and pulled him close, wrapped my arms around his broad, sweaty back, pressed my cheek against the rough nap of his hair, and whispered in his ear, "It's all right, Champ. It's all over now." Sometimes, we all need such touch.

Perhaps you think I'm an apologist for brutality. And perhaps I am. Perhaps I simply want to accept the possibility in each of us. Was the blood or gore or pain of what Tyson did any worse than a head-butt? Tyson, too, was bloodied. Tyson, too, had cut flesh, bruising and swelling. But I understand the difference. There was, in fact, something about putting his mouth to another man's ear, his teeth into another's flesh, something so intimate, the severing so desperate and personal, that it made us recoil and call Tyson an "animal," "psychotic," "savage," or, worse, "inhuman. " I believe what burned us most was the naked humanity, the unfettered vulnerability of what he did. What frightened us was his fear. What disappointed us was his weakness. We are so awful in our love for fighters.

Several years later, in an interview after a warm-up victory against Lou Savarese before a 2001 bout with Lennox Lewis, Tyson seemed juiced with rage, charged full of adrenaline as he barked at reporter Jim Grey a series of mumbled prayers to Muhammad. Grey, already looking ahead to Tyson's bout with Lewis, asked if the fight, which ended in a TKO after only thirty-eight seconds, had been Tyson's shortest fight ever.

Tyson then launched into a monologue wherein he talked at least twice about having to bury his best friend, about how this fight was for the dead friend. He was grieving publicly, painfully, and you could see him focus his grief and his rage on Grey's question and on his role as Tyson the Fighter. You could see the switch flip, a manufactured persona rising to the surface, and Tyson seemed to rev up even more, comparing himself to Alexander the Great, Sonny Liston, and Jack Dempsey, calling his style "impetuous," before eventually threatening to eat Lennox Lewis's children.

Lennox Lewis did not have any children when Tyson said this, but the line, delivered at the end of an outpouring of grief over the loss of his friend, has become, in popular memory, another tag, another mark—further evidence that there is something wrong, something dangerous about Mike Tyson.

It makes me love him even more.

You see, I liked to nibble on my children's ears. Sometimes, I even did it in public. I did it with my lips curled over my teeth so it wouldn't hurt. I nibbled gently like a dog biting a puppy's ears or a monkey grooming his baby. When I carried my daughter on my hip, her round, soft face and ears sat right at mouth-level, and it was all I could do most times not to nuzzle her cheek, kiss her face, and nibble on her ear, rolling the flesh and cartilage between my incisors, the helix of the outer ear between my lips, never my uncovered teeth, never that sharp. It was a way to hold her and keep her close. It was like a kiss, a cuff, and a tug to tell her I love her.

ROUND 6

Tyson suggests a savagery only symbolically contained within the brightly illuminated elevated ring.

—Joyce Carol Oates, "On Mike Tyson"

Sadly, the ear is an often forgotten, underestimated, or disregarded appendage. You're not likely to respond to "I love your ears" as a pickup line. Nonetheless, we depend on the symmetry and subtlety of our ears, their unobtrusive presence as counterweights to each other and aesthetic accessories to our skulls. We notice only when something is wrong. Some of us need them as talismans and triggers for tangential thinking; and nearly all of us know of Van Gogh, the tortured artist and lover who cut off his own ear. We keep this story close as a kind of parable, a lesson or warning, perhaps a story of mythic love—at least, until it begins to mingle with other stories, other parables of severed ears, savagery, mystery, and madness; until it is replaced by Lynch and Tyson, and Travis. For me, there is as much gravity in the story of Van Gogh's desperate act as there is in that image of Holyfield's bloody helix, that small curve of cartilage severed from the rest, and as there is in Lynch's ear cradled in a bed of grass, or Travis's choice to leave Charla's ears alone, a kind of aesthetic and ethical weight that shakes up our measure of the balance between human and inhuman.

We'd like to be able to dismiss Tyson as "animal," a mentally deranged savage who is nothing like you and me. We don't want to admit it, but perhaps it's not that easy to separate us from him. Tyson is no animal, no inhuman fighting machine, no simple boxer; instead, he is perhaps more honestly and innocently human, more vulnerable and real and dangerous than most of us can ever hope to be. To me, Tyson always seemed to be fighting himself with as much savagery as he fought his opponents, and he always seemed sublimely, purely alive in the ring in a way I could never dream to achieve in any context. We want to believe that we're not like this Mike and that we're far from Travis and his brand of savagery, too, but try as we might, we cannot always remove ourselves completely from the urge to bite, to sink our teeth into something substantial, something firm but forgiving, tangible or terrifying or terrestrial—especially when we are at our weakest and most vulnerable, or when our plans collapse and we just want to be close enough to hold someone closer than seems physically possible, to consume that person and keep them inside us forever.

LUPE LINARES

A Living Structure

NUMBER NINETEEN FELL apart unexpectedly when I bit into the world's softest piece of garlic bread. Biting into that bread felt the way I imagined biting into a cloud might—airy, moist. My mouth seemed empty, but I could taste the hint of garlic and the melted mozzarella all around. When my teeth clamped on the hard chunk of something, I spit it out in shock. The little beige rectangle hit my plate, made a sound like two glasses clinking, skidded across it, off the edge, and onto the red-checkered tablecloth.

I could tell it was a tooth, but I hoped it wasn't mine. I ran my tongue across my teeth to check, and sure enough, a third of one of my molars was gone, leaving a cavern in which a morsel of garlic bread lay trapped. The break was clean—a deliberate straight line that made it look as though someone or something had been living inside the tooth, sawing at it until it was weak and could be tapped out of place. I picked up the third of my tooth from the table and held it between my fingers while my tongue probed at what remained. Even though the change happened less than a minute before, I couldn't remember what it felt like not to have that emptiness in my mouth.

Almost a week before my tooth fell apart, my mom called to tell me about my dad's front tooth, which had been damaged in a car accident years before and had been slowly dying ever since. My mom and I were both in the truck when the accident happened. We were driving down one of the narrow, paved roads that winds through Adams County, Pennsylvania, when a car sped around a curve. The car drifted into the middle of the road, and even though my father swerved, he wasn't quick enough. The car slid against the entire length of his truck, shattering the driver's side window and rippling the entire side so that it looked like a crumpled paper bag. I was sitting between my parents holding a can of Dr. Pepper, which didn't spill at all. Still, my mom took the can from me after we stopped and poured it on the ground. "There might be glass in it now," she told me.

We were all fine, but the impact of the accident caused my father's head to jerk forward and his mouth to collide with the steering wheel. There was no visible damage, and if I were him, I would have probably called myself lucky.

Even after the damage started to show years later, he was reluctant to go to the dentist, have the dentist pull the tooth, and then charge him thousands of dollars to construct a new one out of porcelain. He said it would be cheaper in Mexico, and since he needed to visit his mother anyway, he would wait. For years, he tried to avoid biting with the sensitive tooth. When he ate dinner, he would roll up his tortillas and clamp down on them with the right side of his mouth. Neither my mom nor I knew that the tooth was causing him any pain.

The night before my mom called me, she was watching television after dinner. My dad grilled steak that night, and it was tough. After dinner, he went to the bathroom. She thought he was shaving until she heard the scream, and when she made it into the bathroom, my dad stood still with blood dribbling down his chin and landing in heavy drops on the pale, blue sink. He looked into the mirror to inspect the gap in his mouth while still holding the offending tooth between his thumb and index finger. When I asked him why he did it, he said, "It hurt when I ate."

My mom and I both told him to rinse his mouth out three times a day with warm salt water. We had both been through this before and knew what to do, but my dad didn't listen. He said that he was okay, that it didn't hurt anymore.

Had he gone to the dentist, he would have likely been told

that the tooth he had been so desperately trying to keep despite the discomfort it caused him wasn't just a front tooth. It was called number nine, it had a purpose, and it should be missed.

The first time I saw my father without his front tooth was only three days after number nineteen had been reconstructed with silver amalgam. When he smiled at me in the airport, I could see his tongue. My mom told me that he spent the first few days after it was gone holding his hand over his mouth when he talked or smiled. By the time I saw him, he acted like the tooth had never been there in the first place. His mourning period was over. He had accepted his loss and was in no hurry to find a replacement for number nine.

My father has always been a handsome man. In his old driver's license, which I carry with me in my wallet, he is thirty-one. His skin is the color of tanned leather from working outdoors all year long, and his hair is full and black. His moustache, goatee, and eyes are all black, too. He has always had a moustache, though in this picture it is a thick handlebar connecting to the goatee. Over the years, his moustache transformed into a neatly trimmed border that frames his upper lip, and the beard disappeared, never to be seen again. Like his sideburns, the moustache now has streaks of gray in it sometimes. (Occasionally, he says, "I need to paint it," and goes out to buy Just For Men moustache and beard dye.) There are wrinkles in the corners of his eyes now, and his hair has been thinning for years. Still, he doesn't look old yet.

In the driver's license picture, he might look mysterious if he weren't smiling so that you could see his teeth. His teeth were perfect—straight and white without any gaps—until number nine started showing signs that something, somewhere, had gone wrong. It looked nicotine-stained while all of its partners were still white and healthy. The discolored tooth did not diminish my father's good looks, but the gaping hole made him seem incomplete.

My mother is also missing a front tooth, though hers is on the bottom and doesn't show when she smiles. She went to Georgia to visit my grandparents once and came back without it. I was in high school at the time, and I kept asking her how it happened. The hole haunted me, but she wouldn't explain. "I fell down" was all she said. She didn't seem to mind because she had never had nice teeth. She started smoking when she was thirteen, and by the time I came along, they were already beginning to yellow. After a while, I barely noticed the gap. But I did notice how, three years after she lost it, she gained weight and could never get it off.

My mother was thin for only a year of her life. She was twenty-two, and she got that way by eating nothing but canned soup. No breakfast, no lunch. Just soup for dinner. By the time she met my father, she had gained the weight back. She wasn't fat, exactly, but voluptuous with hair that appeared blonder on the ends than it was around the roots, the result of years spent using home hair-dyeing kits. She had long, red fingernails and wore too much makeup. I don't know if my father thought she was beautiful. In all of the pictures I have seen of the two of them before I was born, she has a round face, and she looks shocked—at what, I don't know, but it might be her luck at finding this handsome man who didn't beat her like the two husbands she had before.

Or perhaps that wasn't it at all. Maybe she surprised herself with her own courage—the courage that it took for her to say yes when my dad walked into the bar where she was working one night and said to her, "Apple season starts next week. I'm leaving for Pennsylvania tonight. Are you coming?" She had known him only for a few weeks, and she didn't like him in the beginning. He was cocky, she said. The first night he came to the bar, he had a broken leg. That same night, he got into a fight and won. Later, he asked her for a free beer. She didn't give it to him that night, but eventually they were friends, and she did. When he asked her to leave with him, she didn't think. She just said yes and went back to the trailer where she was staying to pack her things and left the South forever with this Mexican man she knew her mother would never accept.

They drove to Pennsylvania in the car that she had stolen from her second husband not long before. She had been afraid he would kill her, and one night when she thought that he might, she managed to get out of the house and into the car. She drove the forty miles from Bainbridge, Georgia, to some town just across the Florida state line where she had a friend who had a trailer. When my parents got to Pennsylvania, my mom drove that car to a junkyard fifteen miles away. She sold the car for scrap metal and hitchhiked home.

When I think of her this way, she is beautiful to me. The missing tooth; the papery skin that looks so much older than its fifty years; the thick, yellow fingernails that are never painted red anymore—all of this is beautiful because this is what is left of the fearless, unpredictable woman who raised me. These are the marks that remind me of what she has lost, what we have both lost, and only in their presence can I remember her completely.

Now, she cries too much. Her kidneys are failing, her arteries are hardening, and she retains so much fluid that her weight can easily fluctuate fifty pounds in one week. She doesn't drive if she can help it because she doesn't trust her sight, or her hands, or her lungs. She asks for help to do the simplest things—clip her toenails or take off her socks. I call home almost every day to hear her voice, and when she doesn't pick up, my body goes cold. I leave a message, and when she calls back, I can breathe. "Hey, baby," she always says.

Before Dr. Hohlen began to reconstruct number nineteen, he explained that the tooth is a living structure. The pulp contains blood vessels that supply the tooth with nutrients and nerves that allow it to sense hot and cold. The pulp also contains lymph vessels that have the responsibility of carrying white blood cells to the tooth to help it fight off bacteria. Each tooth is independently alive and has its own private system that helps it survive. Unfortunately, each tooth also has to contend with its owner, who has all the power over how it lives or dies.

The acts that can preserve the tooth are so simple and take up an inconsequential portion of the day: brushing carefully so that the bristles of the toothbrush scour the entire surface of each tooth, flossing to dislodge the crumbs that get trapped in between them, and finally rinsing with mouthwash to kill the bacteria. The whole process takes five minutes, but the winding of floss around fingers seems so tedious, and it's easy to miss the backside of the rear molars. Small mistakes that accumulate over time and add up to a loss that we can never forget.

I was home with my parents for a full day before I got used to my dad's missing tooth. The open space was like a window, and I could look into his mouth to see how hard his tongue worked when he talked and how it retracted from his teeth when he smiled. The novelty of witnessing what should have been the private workings of a tongue wore off, and just as it did, the entire left side of my mouth began to ache and didn't stop for six days. My left cheek was swollen, and my jaw wouldn't open more than half an inch.

My dad is a master on the grill and promised to make burgers while I was home. When he finally did, it took forty-five minutes for me to eat one. I, too, began to take my bites using one side of my mouth. Without his front tooth there, my dad couldn't break his habit, so the two of us sat at the kitchen table, him holding his burger to the left side and me holding mine to the right.

When the agony subsided a week later, I was left with a blister the size of a dime on my gums. Because my tooth and jaw were no longer hurting, I convinced myself that the blister was no big deal—it didn't feel much worse than taking a drink of orange juice when there was an open sore in my mouth. I didn't go to the dentist until I made it back home to Nebraska, and even then, I waited three more days.

When Dr. Hohlen looked at the sac on the side of my gums, he said, "That's what I thought," and walked away. In the room where I sat, there was a poster above my head that had a picture of a Shar-Pei wearing gold, wire-rimmed glasses. He looked distinguished, lying on the floor with one paw draped casually over a book. Above him was the phrase "It only hurts when I study." I watched him and poked at the blister with my tongue.

Dr. Hohlen returned and told me that my tooth had abscessed. "What that means," he explained, "is that the pulp got infected, which caused the tooth to die. All of the infection in the tooth had to get out somehow, but because of the filling, it couldn't get out the top. Instead, it ate through your jawbone and leaked into the gum tissue. We'll have to lance that so that we can drain the infection."

Before I left, he told me that the blister was one of the biggest he'd ever seen and that I would have to get a root canal before he could put the crown over my tooth. I walked home with the hole in my gums sutured shut. My tongue moved up and down my dead tooth as though I were petting it.

Number nineteen wasn't the first of my teeth to go. The first one was number two, the back molar on the upper-right side of my jaw, and it fell apart gradually over the course of five months. There was a cavity, I knew, but I was concentrating on finishing my BA and thought that number two could wait. When it fell apart, I was eating a piece of French toast. I felt it dislodge, and the shock of its hardness against my tongue was too much. I threw away the rest of my breakfast but kept the tooth. I wrapped it in a napkin, and when I got back to my dorm room, I tried to fit it back into place just to have one last memory of what it felt like to be whole. When I couldn't, I put it in my desk drawer. I didn't tell anyone except my mom, who never felt much for her teeth anyway and couldn't sympathize.

Five months later, when I finally sat in a dentist chair to have the tooth looked at, I couldn't concentrate. Before I left the house that morning, my mother couldn't breathe. My dad wanted to take her to the hospital himself, but she couldn't move either. She told me to call for an ambulance, but he kept telling me not to, that between the two of us, we could get her to the car. We didn't have health insurance, and I knew the ambulance bill would be expensive, but my mother's breathing sounded like a kazoo. "I can't . . . breathe. I need . . . oxygen," she wheezed. She smoked all of my life, and the sound of her wheezing became normal to me over the years, a cadence that lulled me to sleep at night. This time, though, her face was red.

When I picked up the phone, I had to think about the numbers—911. I learned them in elementary school but never thought I would dial them. I was detached from myself for the whole ordeal, watching from some omniscient place as I told the operator the address and picked up the dog to lock him away in the bathroom before the ambulance arrived. It took three men to get my mother strapped on the gurney, and my dad and I couldn't do anything except stand back and watch. When they left, I looked at my dad.

"What should I do? I'm supposed to go to the dentist in an hour."

"Go. You can't do anything now. I'm going back to work until lunch. We'll call the hospital then."

I went, even though it felt wrong. I didn't listen to music on the way.

At the dentist, I filled out a form, checking no to every allergy. The question that stopped me, that haunts me every time I change dentists, asked: "How do you feel about your teeth?" If I were honest, I would have written that I am so afraid that I might lose them all that I run my tongue over each of them after every meal to monitor any changes and that some days this process can take a full hour. Instead, when no one was looking, I inspected them in the mirror hanging on the wall next to me. They were straight, but there were gaps between them. Number seven was chipped because I hit it against the rim of a thick glass once when I was drunk. They weren't exactly white but the off-white color of my high school's cafeteria walls. I wrote: "I'm not proud of them, but I would prefer to keep them."

When I was finally called back and seated in the dental chair, the new dentist that I found in the phone book didn't talk much. His offices were once a family home, and the room that I waited in had been the kitchen. I knew because one window was shorter than all the rest. When he came into the room, he wore a mask that he didn't take off for the entire visit, and it muffled his voice when he asked how I was doing. He walked out of the room before I could say anything. I was grateful. How could I have told him that I was afraid my mother was dying? And how could I have not?

When the dentist came back in, he told me to open my mouth and scraped at what was left of my tooth so violently that a few more chunks flew off and got caught on my cheek. He looked straight in my eyes when he said, "Well, we can save it for you if you want. It'll cost about fifteen hundred dollars. Or we could just pull it for a hundred. It's up to you, but I would rather pull it."

"I need a minute," I told him. "I don't even know how much of it's left after that."

He left to give me time to think. I knew the answer, but I couldn't let go so quickly. When he came back, I said that I wanted it pulled. I didn't mean that I wanted it pulled at that moment, but he told me he had time. There was nothing I could say—my mother in the hospital, my tooth in pieces. There was no time to say goodbye, and before I had a chance to run my tongue over number two one last time, my mouth was numb with Novocain.

The dentist told me to keep my head still, but his jerk was too strong. My head lunged violently downward with each tug, though my mouth felt only the pressure of his pull and the final give of the roots. The tooth came out in two large pieces, and I made the mistake of looking over at the bloodied roots lying on top of a tray that had been covered with a Snoopy paper towel.

He packed my gums with gauze, and before he sent me to the front room to pay my bill he said, "You would be an excellent candidate for veneers, if you're interested."

Besides not being able to breathe, my mom was fine. I visited her in the hospital the same day on my way to work. I had never seen her with an oxygen tube running under her nose before, and the presence of the thing accentuated the absences I felt but couldn't articulate. When I left the hospital, I sat in my car with my shoulders heaving. I didn't cry, though I performed the motions of it. I didn't know what there was to cry about, but I knew that my life had changed over the course of a few hours. I knew that I had lost something that would never come back to me, but I couldn't say, still can't say, what it was.

I knew only that whatever I lost was replaced by fear and longing. Though I knew that everyone had to die, I hadn't realized that we are always dying, that we die in pieces, and that once those pieces are gone, only absence remains.

When I got to work that afternoon, I changed the bloodied gauze packed against my gums. The bleeding didn't stop all evening, and I couldn't eat the grapes or the peanut butter sandwich that I brought for my dinner. The grapes were lost anyway, rancid from sitting in the sun while I visited my mom.

I drove home weak from blood loss and not eating. My dad got home from his truck-driving job just minutes before I pulled in. He had put in a whole day in the orchard, visited my mom in the hospital, and hauled a load of pulpwood to Westminster, Maryland.

"You look tired, m'hija," he said.

He fried beans for me because they were soft, and I wouldn't have to chew.

As I sat in the dental chair waiting for my brand new crown, I thought about all I had been through with number nineteen. The first time a dentist worked on it, I was thirteen. I was getting cavities filled for the first time in my life, though not because I had never had any. My mother, who had always been afraid of the dentist, couldn't bear to take me.

At that time, my dentist was Dr. McBeth, a short lady with a receding hairline and large breasts. Though I had at least eight cavities filled within a few months of each other, she had never said a word to me. Her assistant had done all the talking, telling me to open wide for the Novocain. "There might be a pinch," she said. I closed my eyes tight. When Dr. McBeth finally started working on number nineteen, she pushed her breasts into the side of my face. They were warm and spongy against my cheek.

After Dr. McBeth was done, there were two fillings on number nineteen—one on the side and the other deep into the center of the tooth. The natural tooth around that filling must have weakened, causing a third of it to break off into my garlic bread more than twelve years later. That break would lead to a reconstruction, which would lead to an infection, then to one failed attempt at a root canal, and a successful one. A month later, there would also be a temporary crown, and finally, it would all culminate in the permanent crown that had been carefully molded and crafted out of porcelain to feel and look just like the old number nineteen.

I was in the room for only a few minutes before Dr. Hohlen walked in. Because the tooth was already dead, he didn't give me any Novocain. I kept my eyes open throughout the whole process. He pried off the plastic crown and scraped off the temporary cement used to keep it in place. He snapped the porcelain crown over the tooth several times, occasionally popping it off with the help of dental floss so that he could sand away a millimeter of porcelain on either side. When he was done and the crown was finally in place, he told me that it looked great.

"Thank you," I said. I looked slowly around the office before I left to take it all in. I would miss it—the sterile yet rubbery smell, the fluorescent light that glared into my eyes, and the set of false teeth with clear gums and a jumbo toothbrush stuck between its jaws.

I would also miss Dr. Hohlen, who said before I left, "You're a really good patient."

Over the three months I spent in and out of the dental chair, I learned how to hold my tongue flat against the floor of my mouth to give the dentist and his assistant full access to all of my teeth. When the time came to clamp them together and tap, tap, tap the red paper so that he could see where he needed to reshape the cusps to fit my bite, my jaw lined up straight every time. I was polite, and asked questions, and said thank you, even when the news wasn't good. I was composed, though I was never in control.

On my way out of the office, I stopped in the bathroom to inspect the new addition. The old number nineteen was filled with silver amalgam, but this new one was pristine. In my mouth, it looked like the only brand new car in a lot full of used ones. I ran my tongue around the edges of it. The porcelain was smooth, and though I knew I would always mourn my loss, I couldn't help but admire the craftsmanship of its replacement.

SARAH ROSE ETTER

The Spine

X.

A year after the Russian man splits you open with a scalpel,
you will look back at a desert made of crushed painkillers, the
white chalk of the pills still stuck between your toes, the world
still covered in that powder, that dust still clinging to your hair,
your face, your skin. In dreams, this is what you choke on. In
dreams, it is how you die that wakes you up.

X.

A spine is a column that holds up a house of skin. A baby is a
memory of a certain point in time. A spine doesn't support a
baby at first. At first, the limbs grow, the muscles become more
like muscles, then there is a crawling, then a standing, then a
walking. That is when a spine establishes itself, proves its value.

Once a baby becomes a woman, she is held up by a spine. A
spine allows a woman twenty-nine years of freedoms: walking,
bending, running, fucking. A spine or a baby can both collapse,
and with them the skin falls inward. This is how motion quits,
how the world stops moving, how a body becomes paralyzed.

Without a spine, there is no action. Without a spine, there is
only lying on carpets, staring at the same blank ceiling, the
heart inside the chest a two-headed fox furred red, two mouths
gnawing through the insides, trying to escape the white ribs,
hungry for air, movement. The definition of hell is an absence of
air, of sleep, of movement, the limbs becoming fat off paralysis.

Doctors are men with fingers that press parts. Pain is a feeling with a scale to it. A slit above the spine is a type of release. Life is a series of surgeries. Some involve knives, others involve silence.

X.

Here is the nightmare: A row of sixteen doctors in white coats. Your body is unopened on a table. You are in a paper gown. Beneath your skin, your blood is lava, so hot the pain makes your body translucent, the ache is so great you become a light, glowing lamp-like, your eyes scattered in different directions from the pain.

One by one, the doctors approach your wretched body, prone on the metal. One by one, the doctors look into your face, into your scattered eyes, and say their scripts:

We can't find anything wrong.
You are fine.
Stop asking for pills.

All around you, the chalk floats down soft, snow dusting the heat of your skin. The powder of the pills builds and builds and builds, covering the color of the lava blaring through your veins. The dust creates small mountains on your nose, your elbows, your legs, from above you look to be coated in powdered sugar. But the throb is still there, the heat of the lava only dormant for the false winter.

All around you, you have quit your life. You've moved out of the city, back into your parents' house. You've stopped going to work. You've started to lose control of your bowels. Six times a day, on average, pain rockets through your spine so sharp that you leave your body, pain beyond pain, pain beyond four letters, pain there is no language for, and you return to your body wretched on the floor.

X.

A spine is a new home for needles, which navigate like hurt rockets through skin, sink through the flame to create a new pressure that erupts up the vertebrae. After this, there is silence in the bones, a quiet for days. This is when a two-headed fox goes to sleep or becomes more violent, chews through even more viscera to escape. This is when every hue takes on a deep calmness, or else a hand itches to write a postcard from a hell pit.

A doctor is a man who can be as wrong as a spine. A nurse is a woman who holds up a wrong man. A body is a piece of meat illuminated on large screens. A father is a man who is at a loss for words. A mother is a lap for your spinning head. Pain is a sensation that wrenches the body when it is not being blocked. A mother is a woman who will stroke your hair when you wish for death out loud.

A house is made of walls of skin. In the attic, the brain is a triangle. Vertebrae make a white ladder from skin to that shape. The most important part of living is protection from damage, the ability to climb. Everywhere, there is danger. Everywhere, there are threats to flesh. Outside of the house of skin and brain and spine there is a mouth with a tongue. Lips are a door. A tongue is a wecome mat. We disagree about the eyes, whether they are windows. There is no way to prove this one way or the other.

X.

The Russian man is bald with crossing wires above the scalp. A scalp can be a chessboard. The Russian man is a king of white cloth that bends over both glowing rays and human bodies. The Russian man determines the problem with the column in the flesh house. The Russian man says your spine has a herniated disc worse than he's seen in thirty years. The Russian man cannot believe you can walk. The Russian man slides his thumb into your ass to determine how much rectal function you have lost over these three months.

He says:

"This is very simple. I cut you open and slide muscles to the left. I tie muscles with neat bow. Then, I cut the spine. Then, I untie bow and slide muscles back."

Everything inside of you is a curtain. Everything can be slid and tied to reveal more. You are a canyon of organs, bones, fat deposits, possible tumors, breast tissue, ligaments, white and red viscera, and then, finally, beneath all of that, the spine, which is your main river only very still, very hard, made of bone.

The only man you have to love is the Russian man. No other man is as concerned with you, as worried. No other man frets over your form, presses his fingers against you, his fine knives. A man has never looked upon your body with such sharp eyes. The Russian man has memorized the freckles constellating on your back, the slope of your ass cresting up out of your spine, down to your legs, the landscape of your whole skin. If nothing else, you have this.

Is the Russian man a father? It is hard to know. You cannot imagine his accent at baseball games or buying small hamburgers at Wendy's. But there is that circle of silver on the certain finger, which means there may be all of that and more, the baseball games and hamburgers, the nights in front of the television after the house has gone to sleep, tears collating at the edges of the eyes during an infomercial, when he realizes nothing will solve the lonely puzzle in his chest.

The Russian man speaks in timelines. The Russian man details a likely path of recovery. By timeline, we mean the hours it will take for the skin to mend itself. By timeline, we mean the minutes it will take for the house to make progress. By timeline, we mean the total milliseconds required for cells to regenerate over wounds.

A spine is an unclasped necklace of bone and disc. A disc is a circle of fiber. When a disc herniates in the lower back, it presses on a bundle of nerves at the base of the spine. When the bundle of nerves at the base of the spine is pressed, everything is lost: feeling, control, motor function. When the disc touches the sciatic nerve, it torches pain through the body. So much pain, the Russian man says, that you could have six children with no painkillers, that you would laugh at it now.

The Russian man says it best, says it simply; when he says it you fall even more madly in love with him, with the promise of the removal of lava from your veins.

"Inside, the disc looks like shards of crystal in blood. I remove shards."

X.

The night before it happens, you are swimming in painkillers in patches sucking at your skin to deliver relief, fog. The night before the Russian man splits you open, you picture him inside of you, his hands deft, sliding small, clear fragments from your flesh, your body just a skin bag of shattered glass.

Before the Russian man there were other men who touched you. They had names and faces and an order. The last one before the incision was the tallest. At the time, that felt important. At the time, that made him a prize.

X.

Unconsciousness happens when a man needs to create a hole in the flesh to enter. A crowd will gather around, all in white, the inverse of a wedding, unless the wedding gown is your nudity. This is a day for no makeup. This is a day to enter false sleep on your back. While you are gone, while you are in a numb landscape, you will be moved. What happens to your limbs? Does the crowd struggle to lift you and put you in different positions, so the Russian can access your spine? This is a complex question that gets at the weight of your form.

You don't know where you go while this is happening. This is a point in time during which you stop existing. There is no accounting for these hours, for the time of the scalpels and masks. You are sure, above you, that there are eyes only. You are sure, only, that this is a medical masquerade.

Anesthesia is a form of becoming only a body. At first, the needle makes it so that only your spine disappears, all of your specific nerves blocked, the pain still there but covered by a kind ether. For hours or decades, you are held up and held down by nothing, your body columnless, a house of skin floating, the walls rippling and shifting when the wind blows.

A loss of consciousness can be reversed. A nightmare can chew on your heart for days. Spend one month with your body always parallel. This is how muscles mend. On special days, your father takes you to the pool. On those days, you can lie down in the grass, in your swimsuit, parallel to the pool, your incision healing under a layer of dry cotton, under the sun. On those days, you are forbidden to enter the water, which looks more gem-like than any day before.

X.

You can split your life into two periods of time: before the Russian man and after the Russian man. No matter how you describe what the Russian man did to you or the aftermath of that slicing, no one seems to understand. You can describe, at length, the pattern the stitches made up your spine. You can go on about the way your head became a permanent moon over the toilet, the contents of your stomach shooting like white rays through black night into the clear sea below.

A scar is a ghost with cells. At a desk, on weekdays, a human body is a vessel for slouching. A spine is a curved stick of chalk.

The buildings all have the windows you are imagining. The people are all wearing the clothes that you wear when you go to work. They are driving their cars, drinking their coffees, wearing their shoulder pads, too. Beneath your clothes and on your body is a thin line above the vertebra, a marking where he entered you, the proof of stitches after he left.

The bathroom is beige and navy blue. Inside of it, you take your shirt off. In the mirror, you look at your mirror body, everything is a little silver, light bouncing around. You turn around and look over your shoulder at your mirror back and you see it for the first time, you see where the pain came from. The slit at the bottom of your spine is parted a bit, the stitches invisible, you can see down through your skin, all the way to the deep flesh, that dark, dark red.

The Russian man checks your healing. You're in the paper gown in his office again, the warmth of his body familiar now. He touches your back, the incision, the muscles. He redresses your wound after he touches you, cleans his hands with sanitizer. At other times, before him, you let insane men touch you and try to heal you. One man jabbed your legs with toothpicks, another one you let inject your back full of saline to dull the pain, and the whole time, as he slid the needles into you, he kept saying Here comes the sugar water, here comes the sugar water.

X.

At times, they will make a temporary corpse out of you. At times, there will be root canals and bad organs and wilting heart veins. At times, you are going to need to be gone from yourself so you can be worked on, so the red machinery of you can run well oiled again, so you can be fixed. Understand, you are just a soft, wet robot with bone for metal.

Nerves are inside of you and they are red. Also inside of you is a splay of deep-viscera black, that horrible red that means death, that vital color. A scar is forever, a way to define a person, a way to find a certain body in a certain crowd. At some point, you can throw away all of the old feelings, force them into a mental fire, burn everything that mattered. At some point, you can start all over again. At some point, it will all be white.

An anniversary is a celebration of a death day. Did the Russian man have small scalpels for fingers or did he have real human hands? You are always remembering him wrong; you can't remember what his eyes looked like; there is a haze around him or a halo so bright it is hard to see anything but the top of his head, bent over you, pressing, pressing, always the pressure of him.

There are anniversaries that don't involve cake. There is a date on the calendar that will rip your heart out with its hands. Isn't that what an anniversary is? On August 22, don't look for the Russian man. He won't show up. Stop picturing him at the front door of your apartment, balloons in hand, a silly hat, that cake, with your name on it in blue.

It takes one to three years for nerves to heal after trauma, if they ever heal at all. A brain is a control board for a body. When a brain is in sustained pain for long periods of time, it rewires itself to cope. It floods itself to sustain. It changes. Studies say the brain can change back, but isn't that always a question? Has your brain changed? It has. Will it change back? Will you know if it does? Who are you now?

A body is a flesh house. A body is holding you. A body is a flimsy thing and all around us are dangers. Most anything can pierce skin, most objects or accidents can draw blood. An unbleeding body is a blessing, practically a miracle. If all of your limbs are intact, be very careful. If all of your limbs are intact, you are temporary porcelain. If all of your limbs are intact, spend the next hours, minutes, seconds waiting for the shattering to happen. The next shattering is always coming.

KAITLYN TEER

Drawing a Breath

I.

IT STARTS AND ends with a breath.

The shock of air on a newborn's cheek, the cold kiss of it. This is what beckons a first breath. It isn't born of need, not a hunger for air nor scarcity that compels us to fill our lungs for the first time. Rather, drawing a breath is our natural response to the abundance of air. Air stirs the fine hairs on infant bodies, sways lanugo like seaweed underwater. A gale-force rush of stimuli triggers neurons, which spark in the brain. Unfamiliar waves of sound and light roll in: vibrations resonate within the ear canal, and brightness flashes upon the retina in fuzzy shapes.

So it happens upon arrival, within the first ten seconds of extrauterine experience. It seems our first breath holds wonder. More an involuntary gasp of surprise than of survival. Upon encountering the world, the most fitting response is to open our mouths and take it in. We expand as it fills us. This initial inhalation is a moment that is unrepeatable: for the first time lungs trade fluid for air, an exchange of gases, and then, the umbilical cord with its oxygen-rich blood is cut.

My friend, who is a midwife, stands in her kitchen making pizza and talking about the first breaths she has witnessed. Her hands hold the fixed ends of a rolling pin as she leans over the counter, working the dough. She tells me about the small, slippery bodies she catches. She holds these curled newborns up to their mothers' chests and waits for the shallow expansion and contraction of tentative lungs under her palms.

She pauses and smiles as her toddler interrupts our conversation. He chases a toy across the tiled kitchen floor while exclaiming, "Woah." He says this many times, always with the same ecstatic appreciation. He is amazed that a toy man can sit in a toy car, amazed at the pizza pan in the oven, amazed at his own reflection as he presses his hands, lips, and nose against the chilled glass of the back door, leaving smudgy fingerprints in fog.

You can measure a lifetime in breaths. An eighty-year-old: 672,768,000 breaths. A fifty-five-year-old: 420,480,000 breaths. A twenty-five-year-old: 210,240,000 breaths. In and out, impossible to count each one, and yet, not one of them is like the first breath.

"I once took a nine-hour seminar on infant resuscitation," my friend says. "Not just the mechanics of it, but to explore the meaning of it, the significance."

She tells me what happens when a baby is born but doesn't breathe, tells me of the stillness in the room except for the effort to bring into being the one small body whose chest has yet to rise and fall.

She tries to tell me what it's like to breathe in and out, offering air to airless lungs. This giving of breath, to breathe for one who has never known air, is unlike any other resuscitation, more a suscitation, a rousing to life.

II.

In the old-growth forest, the nearness of aged beings silences me. It is the quietude of Douglas firs, hush of hundreds of years of respiration. I sense in the air the scent of reciprocity, stomatal openings receiving carbon dioxide and delivering oxygen in return.

My parents have traveled across the country to visit us and they walk with my husband and me under a canopy of firs as we hike toward Fragrance Lake. Mottled light covers the chipped path before us, carved into a carpet of vigorous sword ferns. All around us, moss makes its patient creep across bark and stone. The first mile is a steep incline. My father pauses occasionally to unclip his cell phone from his belt and take a photo, to study the globed web of an orb-weaver spider or the exposed root system of a fallen tree. He lags behind us, taking his time, smiling up at us during switchbacks. When he stops again to retie the laces of his sneakers, I fall back to walk with him. I overhear his ragged breathing and listen for what each rasp might mean.

I don't mention my father's shortness of breath, but together we decide against finishing the hike. Instead, we follow the spur trail to a lookout point, which offers a view of Samish Bay and the San Juan Islands. We catch our breath, gather it up from the ocean, from the plants below the water's surface which exhale half the world's oxygen.

There at the lookout, with the islands rising and falling below us, my mother and I pose for a photo. I turn around to ask my father if he's ready to return to the trailhead, and I'm surprised to find that he's climbed the trunk of a tree growing out of the ground slantwise in order to get a better view of the bay. He is fifty-five, and I feel as if I am encountering him middle-aged for the first time, an amalgamation of the boy he once was and the aged man he is becoming.

Later, sitting in sunlight and drinking beer together, warmth and easy laughter makes us peers until a chill forces us to notice the passage of time. I detect my father's breath made visible against the deep blue air of late summer.

III.

I feel uncomfortable looking at it: this proof of how long we have understood the breath as inextricably linked to life. An ancient amulet of stone, carved into the shape of windpipe and lungs. It was once intended for the afterlife, to protect the organs it represents, but the amulet was excavated from a tomb, catalogued, and photographed.

The Brooklyn Museum obtained the amulet from the family of Charles Edwin Wilbour, a nineteenth-century American Egyptologist. His children donated it to the museum, along with the rest of his collection, as a memorial to their father. His death occasioned the changing of hands.

I look at its image on the screen and imagine holding it in my hand like a marble chess piece from the set on which my father taught me to play, or as one of the many rocks I collected as a child and stored in shoeboxes stacked in my closet and under my bed. I feel the invisible weight of it, a reminder that this amulet is not mine to hold, meant instead for the dark places of decay.

The amulet belonged to somebody once before. Others handled it first: the hands of a craftsman carved the figure from stone using hammer and chisel and file, scattering chipped flakes. The hands of an embalmer made an incision, extracted organs and placed them in canopic jars, returned the heart to its perfumed cavity, tucked the amulet between cloth bandages, which encircled the chest of the corpse. Thousands of years later, a disturbed tomb. This palm-sized stone, which held the hope of breath resurrected, emerged cold and dusty.

Now, museum curators lean over the amulet on display, their focused breath falling upon the glass case.

IV.

When it's autumn, my sister calls me on a Monday before work. She assures me that everything is fine, but she also tells me that our father is at the emergency room. He thinks he's having a heart attack. The symptoms are consistent; she lists the pain in his arm, high blood pressure, and shortness of breath. But he is stable, so my mother and sister persuade me that I do not need to fly home.

An angiogram reveals that my father's arteries do not require stents, though his heart shows signs of strain. Most alarming, is that the pressure in his lungs measures four times higher than normal. So, the diagnosis is severe pulmonary hypertension. After several days of scans and medical imaging, his doctors finally sight the cause. His legs and lungs contain many clots. A revised diagnosis: chronic thromboembolism pulmonary hypertension, a disease that often goes unseen for years, until the patient's clotted breathing becomes too difficult to ignore.

My aunt sends me a text message with a photo of my dad sitting up, smiling, and eating a meal from his tray. I notice that he hasn't shaved for several days, and the shadow of gray stubble makes me uneasy. The plastic tubing of his pronged nasal cannula delivers a controlled flow of oxygen, supplementing what his body cannot easily accept from the air.

I picture my father sitting on a hospital bed fidgeting with the thin fabric of his gown. He feels the cold tap of a stethoscope against his back and a hand on his shoulder.

Sit up straight, please. Now breathe deeply.

I imagine the many people with disinfected hands who have eavesdropped on my father's breathing, sounded the cavity of his chest, read the dark radiographs and scans of his lungs illuminated on a wall, drawn his blood, administered oxygen. I think, too, of the people who have come to care for the parts of him that cannot be touched or imaged, the family and friends who sit by the hospital window watching baseball on the television, who bring flowers and coffee, who pray for him, who text me to say that God has a plan.

v.

It was dust gathered in hands and a divine breath, according to scripture, that made a living soul. The soul: I cannot say exactly what it is, what the word means inside of a body, the low vibration the word makes when spoken. But when I parse its origins,

I find the confluence of soul and breath, the ethereal and the palpable: *nepes,* the Hebrew word for soul, means a breathing being; *pneuma,* Greek for spirit, also means breath.

The impetus of dissection was soul-searching. Supplied by grave robbers, Leonardo da Vinci opened up decaying bodies, even those of pregnant women and stillborn children, to search for the coordinates of the soul, a quest for the fixed position of the unseen. The same hands that painted *The Last Supper,* applying careful brushstrokes of tempera to a convent's wall, also applied his knife to cadavers.

He was drawn to perspective, to realistic representations of parts and their functions. Once he exhausted his study of machines and the mechanizations of pulleys, levers, and tools, he turned his eye on the human body, to make the incision and peer within. He drew the mechanism of the breath as if it were made of metal and wires.

"From the heart, impurities or 'sooty vapors' are carried back to the lung by way of the pulmonary artery," he wrote, "to be exhaled to the outer air."

His sketch of lungs reveals a steady hand. He mapped the snaking path of arteries, the framework of bones, the folds of intestines, the webbed lobes of lungs. He could not place the soul, to draw its interstices and attachments, to make it visible.

VI.

If I were to take up pencil and sketch my father's airways, to trace the microscopic surfaces of my father's tracheobronchial tree, the task would require fifteen hundred miles of graphite lines. And still, that would be shorter than the distance of flying home if his condition worsened.

My dad calls to tell me he will need to travel to San Diego for a highly specialized surgery, called a pulmonary thromboendarterectomy, to remove the pervasive scar-like obstructions from his arteries. My father is elated by the alternative to a lung transplant and the possibility of making a full recovery.

Surgery is an option I investigated weeks prior. A delicate operation, I watched a video of it on the surgical center's website, studying their hands as surgeons extract clots, fleshy tissues that they arrange on a blue surgical towel, lined up like ginger roots plucked from the garden. While they work, controlled hypothermia protects the body's tissues and organs from damage; the patient is frozen to time, a cessation of brain activity and circulation, a life suspended.

Twice during surgery, they turn off the heart-lung machine, for up to twenty minutes, rendering the surgical field bloodless and the patient clinically dead. But the surgeon's hands have the power to flip the switch, restore the patient's circulation, warm his body, and bring him back. I don't know what to make of these pauses.

My dad calls and we don't discuss these details. He tells me a story that made him laugh earlier that day: "When I was a kid, we'd drive down to visit family in southern Illinois. 'Down home,' my mother called it. She'd bring six, maybe seven, packs of cigarettes. I'd swipe one from the car, and me and my cousins would hide out in the henhouse at Uncle Ed's place."

They smoked through the pack together, emerging from the dense, hot shack, smelling of chicken shit and smoke.

I keep quiet about the standstill, the implications of suspended animation. I want to avoid puncturing his optimism, imagine it as if he's blowing a bubble, air expanding within a slick orb, fraught with all the usual perils of surface tension. Picture it growing larger and wobblier. Picture it floating away.

VII.

I watch as a performance artist brings a self-portrait to his face, so that black scribbled lines look out at his audience. He purses his lips and inhales deeply, while releasing his grasp on the drawing paper. The force of his drawn breath holds the portrait to his face, freezes it there until his next exhalation. The camera zooms in. He sets it down and inhales another portrait to his face.

He inhales one hundred self-portraits. He releases his breath, and one hundred times a sketched representation of his face falls, revealing the face of skin and muscle and bone underneath. He is middle-aged and graying, and his face changes throughout the strenuous performance. With each breath, a loss of water, perhaps more than two fluid ounces under the strain. I worry he will hyperventilate.

Sheet after sheet of paper falls to the floor.

VIII.

I have a premature sense of loss, can already feel the absence in the hospital room as attendants wheel my father's hospital bed away, the strange anticipation, knowing that he will die twice before I see him again. I remember when his mother stopped breathing in a hospice bed. We prayed after she left us; he committed her spirit into the hands of God.

I know that I don't want for him to be bereft of his last breath, cold, split open, and alone, a machine breathing for him.

I understand that after the resurrection, Thomas wanted proof, needed to see the pierced side, feel the flesh and blood with his own hand in order to believe, a kind of closure. And I'm surprised by my want for proof of suspended animation, to see the blood suctioned from my father's body, the heart-lung machine turned off, the circulation's flat line. The need to be in the operating room when he gives up his breath, when his breath returns.

Before, I did not know this desire of mine: to be with the ones I love when they stop breathing, not to leave them lonely.

IX.

Most adults can hold their breath for up to a minute before excess carbon dioxide accrues in the bloodstream, turning blood acidic. But a free diver presses on from the diaphragm, drawing a deeper breath, holding it longer through uncomfortable contractions, following the rope line down a cave's open mouth.

The sunlight filters through seawater and strands of bull kelp as a wetsuit-clad figure walks underwater toward the cave's edge. Above, other dark figures float nearby, descending and ascending an oceanic ladder. Below, the diver's bare feet send a cloudy spill of sand over the edge. A moment of preparation, and then the dive into the sink hole, arms spread, chest first.

He draws his arms back toward his body, streamlined, goggles and black cap leading his descent. As the dark of his wetsuit disappears into the chasm, only the bright spots of his bare hands and feet remain visible, until these too fade from sight.

Everyone watches for his return. The record-holding free dive is twenty-two minutes. He swims down more than three hundred feet before making a turn for the surface. At these underwater depths, human hearts beat slower, just fourteen times per minute, one-third the heart rate of a coma patient. Half the divers who attempt to reach this depth return to the surface unconscious. Yet expert free divers retain control in deep water by training their mammalian dive reflex. Still, they must practice rising. In order to avoid a loss of motor control or the danger of shallow-water blackouts, they practice releasing pressure, expelling air during their ascension, the art of returning to the surface.

x.

My parents pick us up at the airport for Christmas. At their home, I notice on the bureau by the door an arrangement of flowers from the hospital, a spray of baby's breath and four small purple flowers. The tiny white blooms branch out on spindly stems. They are browning, becoming brittle and delicate.

Winter in the Midwest and all deciduous trees look bronchial to me, but we pretend our way through the holidays. My father's illness is an undercurrent, and we are all treading, waiting for the phone call with the surgery date. When he prays to bless the holiday meal, he leaves the dining table. My mother follows him to the bedroom. My husband, sister, and I begin to eat.

XI.

The lung float test: Lungs are the only organs, when extracted from a body, that float. Forensic examiners once looked to the lungs during an autopsy to determine when an infant died. According to this theory, the lungs of stillborn babies, whose alveoli have yet to bloom with air, never having experienced its abundance, sink. The lungs of those who have partaken of air, drawing it in and absorbing the wonder of it, float.

Lungs are named for their lightness, known in Old English as *lunge,* the light organ.

XII.

While waiting for a surgery date, the pressures in my father's lungs subside after months of rest and a regimen of anticoagulants despite the remaining clots, leaving us all to grapple with what his doctors cannot explain, to accept what many call a miracle.

"Enjoy your life," his specialist said. "See you in a couple months."

When I find out my father will not need surgery, I exhale. It's a slow rise to the surface, but I release the pressure that's been building inside me, undetected.

Who can say what prompts a final breath? Lifted by the buoyancy of lungs, to ascend when it is finished, to release what we've taken in of the world, exhaling fully and finally, and away.

KATHERINE E. STANDEFER

Shock to the Heart, Or: A Primer on the Practical Applications of Electricity

Shock to the Heart.
In the first instant, my hands became claws. Paralyzed, a red-hot whip tearing through my back: *Did somebody kick in my spine?* And then I knew. And I was screaming.
"There's no way you wouldn't scream if you felt it," my sister had said.

A Practical Application.
The intentional use of electricity upon the human body dates back at least 122 years, to the electrocution of William Kemmler in 1890 in New York's Auburn Prison. Kemmler—a native of Buffalo, who murdered his common-law wife, Tillie, with a hatchet—was sentenced to death in 1889, narrowly evading death by hanging. His electric chair was first issue, sparkling clean.

A Review of the Literature.
The amount of electricity it takes to kill or damage a human varies considerably, depending on where the current originates and its contact with critical body parts—namely, the heart and the brain. The level of current is the amount of voltage applied divided by the resistance that voltage encounters. According to *The Physics Factbook*, as little as .06 milliamps can cause fatality by stimulating dangerous arrhythmias in the heart. Humans can generally perceive shocks at 1 to 5 milliamps, and 10 milliamps is the level "where pain is sensed." At 100 milliamps, severe muscular contractions occur; at 200, severe burning.

Shock to the Heart.
I was down on my knees in the soccer field grass, facing the backs of houses, where kitchen lights glinted out the windows—dull, far away. The sky was navy and full of cold. After three years, this was it: my internal cardiac defibrillator firing for the first time.

I imagined I must be searing with light. I imagined everyone could see my bones.

When the defibrillator thumped a second time, I knew I would die, lightning-struck, unable to move. Lightning strapped to my heart: *Why aren't I dizzy? Why am I awake?* My chest and arms shrinking, lava-hot with electricity. The moment unfolding endlessly, a maul to the chest, a thousand needles. "Someone call 911!" I screamed.

Glossary: Congenital Long QT Syndrome.
A genetic condition in which the heart's electrical preparation for its next beat (repolarization) can take too long. The delay is exacerbated by various types of physiological stress, most related to the release of epinephrine in the body. This delay, throwing off the rhythm of the heart, can trigger fatal arrhythmias. The condition is typically treated with adrenergic-blocking drugs (beta blockers), which decrease the effects of adrenaline. Those with documented cardiac arrest may also have a cardiac defibrillator implanted in their bodies, to restart the heart if it is inadequately pumping blood.

A Practical Application.
From the beginning, Kemmler's lawyers argued that electrocution was "cruel and unusual punishment." But this was the height of the War of Currents, when Nikola Tesla's new transformer for alternating current (AC) had Thomas Edison (and his business-tycoon backers) scrambling to maintain the use of direct current in public utilities, thus protecting their patent royalties. According to Mark Essig's *Edison and the Electric Chair,* Edison attempted to smear the public image of AC by electrocut-

ing animals with AC in high-profile "studies" and secretly paying Harold P. Brown and Arthur Kennelly to invent the electric chair. When Kemmler's lawyers fought for an alternate method of execution, the prosecution received substantial funding from Edison's friend J. P. Morgan. The electrocution would proceed: a demonstration of the treachery of AC to all.

Lightning Flowers.
A sign of blessing. A curse. God's warning, God's vengeance.
Lightning in the body heals illness. Cures deafness, cures blindness.
Nurtures psychic powers.
From the east, a good omen. From the west, ominous.
Sparks hypersexuality.
I devour the stories I hear. I crave a mythology for the lightning of my own body.

Shock to the Heart.
A third shock—*maybe it's misfiring*—pummeled by hot razor blades—*maybe it's never going to stop*. Realizing I could either breathe or scream, what had I learned? *You must breathe.* "You got this, heart," I whispered fiercely, felt my body solid beneath me, began to pull in air, cold heavy breaths. *I am either alive or dead and I choose which.*
The device did not go off again.
"Can I get someone behind me?" I called out from my knees. "I don't trust myself not to fall." Someone cupped my back immediately, supported me on the ground. The sky my full view. The sharp white field lights. A ring of faces. It may be crazy, but what I smelled was burning.

A Review of the Literature.
Doctors call them "electrical storms"—clusters of shocks, three or more within a twenty-four-hour period. Ten to twenty percent of all internal cardiac defibrillator (ICD) patients experience this within the first two years after implantation. Multiple

shocks generally indicate malfunction—the ICD firing inappropriately—as one shock should be sufficient to disrupt an arrhythmia. According to Dr. Merritt H. Raitt in the *Journal of the American College of Cardiology,* "Multiple shocks are the most frightening for patients, causing them to wonder if the device is really working or if it might even kill them." Even single shocks often cause significant psychological after-effects, including "heightened self-monitoring of bodily functions, increased anxiety, uncertainty, increased dependence, reactive depression, helplessness, and post-traumatic stress disorder."

Lightning Flowers.
In the days after, I bend over my scorched heart. I sleep in fits. I stand before the mirror, eyeing flawless porcelain skin. There are no burns. No lightning flowers spread like pink trees across my breasts. There is no entry point, no exit.

I am both the entry and exit.

Glossary: Terms of Electricity.
If electricity floods my body like a river, voltage measures the pressure of the water. Resistance is the amount the flood is absorbed and diverted. Amperes are the current, the result of pressure diffused—by rocks, riverbanks, skin.

A Practical Application.
On November 11, 2012, at 7:23 pm MST, the St. Jude Atlas II (Model v-268, serial number 375831, 78 grams, implant date 26/OCT/2009) revved its Lithium Silver Vanadium Oxide battery to generate the standard 820 volts of electricity within a thirty-second window. Maybe less. In the first chamber, the capacitor held back the generated voltage until fully amassed. Then it discharged into the center of my heart.

A Review of the Literature.
For lightning-strike victims, the technical name is Lichtenberg figures. These thin, branched burns uncoil from the head, neck,

and shoulders, where a strike is most likely. Fern-like, following patterns of moisture—rain, sweat—they are rose-colored lightning bolts frozen onto the body.

Industrial shocks are different, blisters springing up from the hands and wrists where a person worked a machine or made a repair. The tissue scorches along the entire current, the electricity sizzling a path of least resistance through the body, sometimes leaving dead tissue deep inside that becomes black, gangrenous. Here the electricity runs about 20 to 63 kilovolts, the shock lasting half a second. Then the circuit breakers open. Or the body is thrown.

Shock to the Heart.
Jess tucked a sweatshirt over me. She rubbed my arm, hard. People were on the phone with the dispatcher. They were asking me what it was I had in my chest. They were asking me how old I was.
"How do you feel?" someone asked.
"Clear," I said.

A Practical Application.
On August 6, 1890, Kemmler absorbed 1,000 volts of electricity for seventeen seconds. He was pronounced dead: the first death by electrocution! Only he wasn't. Someone noticed he was still breathing, and another shock of 2,000 volts was required to finish him off. According to witnesses, Kemmler's blood vessels ruptured, the bleeding beneath the skin patching across his face. His hair singed around the electrodes, and the foul smell of burning flesh filled the room so completely that people tried to leave. "They would have done better using an axe," grumbled one witness.

A Review of the Literature.
ICD patients who have been shocked have a higher mortality rate, says Dr. Raitt: the middle layer of heart muscle, the myocardium, can "become damaged." (Does he mean seared?

scarred? torn? blunted?) Or maybe the risk of death comes from the way terror seeps into the body, afterward—quiet and dark, ever present.

Shock to the Heart.

The paramedics, with tanned skin and clear blue eyes, drove their ambulance onto the soccer field and tumbled out. Thick fireman hands slid a blood pressure cuff over my bicep, lifted my shirt to affix electrocardiogram electrodes. One man at a time carefully asked me questions. I liked them. One pricked my finger—"It'll be just a little poke." We spoke a language I'd forgotten I was fluent in, the electricity of the heart, QT interval, arrhythmia. They asked the right questions, about blood sugar and dizziness. When my vitals were good, one of them said, "Do you think you can sit up?" and they supported me from behind and watched my heart on the EKG. Then standing: still good. We all looked at each other very seriously. With no arrhythmia present, the ER could do nothing but waste money. One man gently peeled the adhesive electrodes off my belly. They trusted me to go.

Lightning Flowers.

A fulgurite is a lightning flower in quartzose sand or silica. From the Latin *fulgur,* meaning "thunderbolt," these hollow glass tubes form when lightning with a temperature of at least 1800 degrees Celsius sizzles through the ground, instantly melting silica on conductive surfaces and fusing grains together. A fulgurite is petrified lightning. Its rocky arms twist, fern-like, lightning flowers preserved. Some lace forty, fifty feet into the earth.

Shock to the Heart.

"Do you need us to call someone?" people asked again and again.

"Not yet," I said. My family members all lived thousands of miles away from my home in Tucson, Arizona. There was no need to spread the word while I was still on my back. After I limped off the field, I called my sister in Colorado first because she too speaks the language of electricity. She has been shocked four times: Twice during physical activity—a problem with her settings. Once when the lead wire shifted in her heart. The last, a jolt of electricity that saved her life.

My call caught my sister in bed, reading in a pile of pillows. "Three times in a row," she kept saying. "Three times in a row!" Unlike everyone else in my life, she could actually imagine what this was like. I pictured her in a tank top, blonde hair falling over her shoulders. Twenty-four years old. Probably touching the pale earthworm scar above her left breast. She didn't think she would sleep that night, she told me, snugged into the covers. "I'm suddenly terrified that it's going to go off. I know that's crazy. But I'd forgotten about it, and now I remember again."

"It's so awful," I said.

"Yes," she said. "Now you know what I meant."

A Practical Application.

As it released the electricity that night on the soccer field, the device recorded a heart rate of 175 beats per minute. The device also recorded sinus tachycardia. That is to say: it recorded an exercise-induced elevated heart rate. It did not record ventricular tachycardia, or the torsades de pointes arrhythmia. It did not record ventricular fibrillation, that death-rattle of the heart. It recorded a woman in black long johns and GoreTex trail running shoes, sweating, breathing hard, after a defensive sprint toward the goal.

The device did not record the crisp November night. Not the white lights across the grass, or the dark neighborhoods etched onto an old desert wash. It missed the water bottles of hot tea waiting by the bleachers, the red and gray T-shirts Sharpied with numbers, the guy on the other team who had tripped and was getting up slowly, brushing his thighs.

And the device missed the woman's confidence—carefully knit by three years of its silence—that it had nothing to say.

Lightning Flowers.

If I excavate my body, will I be full of diamond-studded fulgurites? Are my gems pink, like the tissues that burned, or dark brown, something melted? Did the electricity travel according to moisture, according to veins or muscles? Or in some other fashion, a hot random spread? Am I like the ground? The thin soil cover hot and flashing?

A Practical Application.

The device knew only what devices know: whether or not the criteria were being met. The truth was, some criteria were and some were not, and the machine was confused, if machines can be such a thing. Inside the motherboard, the settings said, "If any." So even though there was no arrhythmia—and even though the heart had not rocketed above its set "200 beats per minute for six beats or more" line in the sand—when the dual-chamber Atlas II Model v-268 noted that my heart had been above 170 for more than three minutes, a criterion was satisfied. "If any." Inside the laser-sealed titanium box—no larger than a pager—the capacitor whirled into generation. The second chamber sent the shock.

A Review of the Literature.

Lightning strikes inhabit the body for just a few milliseconds, but sear more than 300 kilovolts. The heart is the most susceptible to harm from electricity. Then the brain.

Anything above 1 amp can cause permanent cellular damage: coagulation necrosis. The dead cell proteins thicken into a gelatinous mass. The architecture of the dead cell is maintained, ghostly white under a microscope.

"Nerves, designed to carry electrical signals, and muscle and blood vessels, because of their high electrolyte and water content, have a low resistance and are good conductors," write University of Illinois at Chicago researchers Mary Ann Cooper and Timothy G. Price in their report "Electrical and Lightning Injuries." "Bone, tendon and fat . . . have a very high resistance and tend to heat up and coagulate rather than transmit current."

Shock to the Heart.

The question, of course, was why today? And not that other day? Why not when I was traveling in West Africa and that man was rattling our windows trying to break in and I was so wound up I couldn't sleep? Why not the night, in the icy Colorado spring air, a lover refused to kiss me? Or the hundreds of times since implantation that I've sprinted up a muddy trail, grunted through rock climbing moves, or carried a pack up the backside of a valley?

A Practical Application.

Traditionally, voltages of around 2,000 have been used in the electric chair, although here, too, there is a range: from 500 on the lower end up to 2,300, which was necessary to end the life of 350-pound Allen Lee "Tiny" Davis in 1999. Reports noted muffled screaming and blood oozing from his mouth and chest. Reports always note the clenched fists, the jerking bodies. Some prisoners' heads burst into flames. Some require such lengthy shocks that the transformer itself burns.

Glossary: Terms of Electricity.

I cannot tell you what a fucking joule is. Or, I can, but I do not understand how an equation translates to chest tissue, blood vessels, shock to the heart. A joule is the energy expended in applying a force of one newton for one meter. A joule is the energy expended to produce a watt of power for one second. A joule is the energy expended in passing an electric current of one ampere through a resistance of one ohm for one second.

Shock to the Heart.

"Why this time?" my sister said. "I can't believe I went to the amusement park. I can't believe I went to the haunted house. I can't believe I forgot."

Lightning Flowers.

Is it I who have shocked myself? Is this bit of metal leaning on my heart, with its fingers plugged into the valves, not a part of me? Doesn't it listen to my every move, kiss the inside of my chest? Is it not somehow me, now, when the muscles stopped complaining years ago, and the scar tissue holds it firmly in place? When it is gently covered in waves of clots, flowers of the body cavity?

And if it is not me, what is it? Palmfuls of strip-mined mountain, knived inside my body? A cache of copper from Congo, a lost bit of motherboard, hidden battery, the laser that sealed it shut?

A Practical Application.

Taser shocks typically last half a second and disrupt the central nervous system, causing "intense pain and muscle contractions." For maximum effectiveness, a Taser should be used on the upper shoulder, upper hip, or below the ribcage. Though the associated current is very low, some sources indicate that a Taser's output can be up to 50 kilovolts, or 50,000 volts, or fifty-eight times the discharge of the Atlas II ICD, or twenty-five to fifty times the voltage used in the electric chair, or one six-thousandth of a bolt of lightning. How much this affects the body depends upon the level of resistance. That is, skin type, moisture, body salinity, shot location, body mass index, and clothing matter, as do the conditions within the weapon circuitry, including the battery and current waveform. And—though the manufacturers insist that Tasers cause no lasting damage, due to a current of just a few milliamps—a 2008 report by Amnesty International documents hundreds of instances in which autopsies suggest that Tasers may have triggered cardiac arrest through repeated or prolonged shocks, or in "vulnerable populations" such as the elderly, people with underlying heart conditions, and those currently under the influence of drugs.

Shock to the Heart.
The next morning, I woke early and called the hospital. The pacemaker clinic could fit me in. I drove myself—somehow the risk of being shocked while driving seemed more tenable than while biking. I am a loner. I live alone. For one morning, this seemed foolish. I shouldn't have worried. I did not have an arrhythmia. The machine was not broken. The settings were just too conservative. "These would be good settings for a fifty-five- or sixty-year-old," the tech joked. I was actually not surprised; the electrophysiologist who had implanted the device, back in Colorado, was fiercely protective of my sister and me. In his waiting room, we were the only people under sixty. She'd passed out in her dorm room. I'd passed out in a parking lot. Death seemed so close then. No one had ever tweaked the settings in the quiet years afterward.

A Review of the Literature.
As the heart rate approaches a trigger zone, the ICD will automatically begin generating electricity. If the actual trigger is not reached, the electricity is dumped, dissipated, put through a resistor and released as heat into the body. The body—talking to a lover or drinking water at halftime—never knows the difference.

A Practical Application.
The idea, of course, is that the ICD will shock you when you're actually dead. (It was known in the 1980s as the "Lazarus device.") If the heart is fibrillating—if it sits quivering on the ribcage letting oxygen waste out of other organs—maybe a nice 820 volts of electricity is just the thing. Say you're lying on the kitchen floor. Or say you, like my sister, are unconscious in bed after the alarm clock startles your poor heart into a frenzy.

The machine goes off and the arrhythmia is disrupted. The heart flatlines and the natural electrical system of the heart kicks back in. Beating resumes.

Who cares, then, about the myocardium? Who cares about trauma and stress? Who cares, as long as you're not dead?

A Review of the Literature.
A recent study published in *The New England Journal of Medicine* found that "simply raising the heart rate at which the device is set to deliver a shock resulted in an 80 percent to 90 percent drop in unnecessary, distressing and painful shocks for heart rhythms that aren't life threatening." Tweaking this setting, researchers said, could reduce a patient's risk of death by as much as 55 percent.

The study was published five days before my shocks.

Lightning Flowers.
Without this metal box—a tiny computer braced into a shell, stuffed with battery, gold wires smaller than human hairs, components glinting up like rows of solar panels—would I be dead? If it has only ever shocked me in error, should I still consider it a savior?

When I held my arm out for the anesthesia—when I opened my chest to the knife—was it just a permission slip I asked for? To live, despite the fact of death? Did I understand I would contain lightning, that I would be cut with current, that I would coagulate, that I could, in fact, still die?

A Practical Application.
"Thirty-five to forty joules is the biggest shock," one of St. Jude Medical's engineers tells me. He says they keep the current small. "If you have a high current, you're going to cook someone." This is because there's no skin, highly resistant, to absorb the flash. No Lichtenberg flowers printing the patterns of rain, sweat. There is just the heart: a direct hit.

A Review of the Literature.
Roy Sullivan's astrological chart is available online, but no one has interpreted it. No one can say why he was struck by

lightning seven times. He holds the Guinness World Record for this—grudgingly. Or he would be grudging, had he not shot himself at age seventy-one out of unrequited love.

A Practical Application.

Today the electric chair is out of commission in every state but Alabama, Florida, South Carolina, Kentucky, and Virginia. Many states have banned the death penalty altogether; in others, electrocution has been ruled "cruel and unusual punishment" because multiple shocks are often required. Lethal injection is considered more humane (although in a recent botched Arizona execution, it took fifteen injections to kill an inmate, who gasped more than six hundred times and took two hours to die). In states like Arkansas and Oklahoma, the electric chair remains in the wings, to be used if preferred by the prisoner or if other methods are found unconstitutional. The United States is the only country in the world to have institutionally used the electric chair, with the exception of the Philippines during American occupation.

Body Memory.

This is not the first time I have been shocked. My body remembers what I cannot: after they corkscrewed the lead wire into the weedy trabeculae at the bottom of my right ventricle—and before they sewed me up—they stopped my heart to make sure the defibrillator would go off properly.

When I woke, there was no telling. Already, doctors had hollowed a cavern in my body, skin tugging over metal, staples stiff in flesh. There was the deep-down ache of the organ taking wire. If my tissues were sore then from electricity, I didn't know it. Maybe neither did they.

A Practical Application.

In 2007, the United Nation's Committee Against Torture ruled that Tasers were a form of torture. According to Human Rights Watch, the Committee Against Torture was particularly concerned that "there is often no physical sign on the body to show the effects," making it hard to tell when a Taser is being regularly or inappropriately used upon a person, particularly in prisons. They also cite concern about the level of electricity.

Glossary: Beta-adrenergic Blocking Agents.
Small white pills. "Beta blockers," as they are known, impede the effects of adrenaline on the heart, keeping heart rate and blood pressure low. For this reason, they significantly decrease the risk of certain arrhythmias, including those associated with Congenital Long QT Syndrome.

Reasons a person may transition off beta blockers include: Desire to listen to the physical nuances of the heart. Desire to spend kidney energy on things other than metabolizing beta blockers. Desire to no longer experience throat constriction and a masking of low blood sugar symptoms. Desire to rise in the morning feeling clear. Desire to operate at high altitudes. Desire to sprint across a soccer field, top speed, without getting dizzy under that star-spackled sky.

Particularly: Desire to honestly reflect updated agreement with death.

A Review of the Literature.
The chance of getting hit by lightning seven times is roughly 22 septillion to 1. That is 22,000,000,000,000,000,000,000,000 to 1. It may have been lower for Sullivan, who worked as a park ranger in Shenandoah National Park, where his risk of contact with lightning storms was higher due to location and task.

Sullivan was struck by lightning in a lookout tower. Then in his truck. Then in his front yard. Then inside a ranger station. In this instance, his head caught fire. Failing to smother the flames with his jacket, he rushed into the bathroom, where his head did not fit beneath the nozzle. The fire was finally extinguished with a wet, slapping towel. After this, he always carried a can of water in his car. He feared death. He assumed there was a dark force after him, and if a storm began while he was driving, he would pull over and curl up in the front seat.

The fifth time Sullivan was struck by lightning, he sighted a cloud boiling up in the park and drove away quickly. He reported that it followed him. When he decided he'd adequately eluded the cloud, he finally left the truck—and was struck by lightning. It lit his head on fire. Still conscious, he belly-crawled back to the truck and poured his can of water on his head.

Repeat this story for the sixth hit: see cloud, outrun, cloud wins.

Shock to the Heart.
I may as well have had no legs, no face, no ribcage. I may as well have been just a receptacle for a defibrillator. In the hospital's pacemaker clinic, the tech clicked at the computer and barely looked at me. He had a gruff voice and a red bandana over gray hair. He did not ask how I was. When I told him I no longer took my medication, he rolled his eyes, shook his head. "Take your medication," he said, "or stop working out. Simple."

The doctor was next, and she spoke looking at the screen. Within her brain existed pages of textbooks, endless studies. I could be grateful for this: she pushed the magic buttons to raise my settings. But when she was done with the computer, she did not turn to me. She did not ask how I was. She, also, did not ask why I was off my medication. She simply picked up her prescription pad, wrote a new one, and handed it to me. "You should probably make an appointment to get established sometime," she said. And walked out.

A Review of the Literature.
The literature does not show who it was that Roy Sullivan loved. The literature does not show the make and model of his truck, whether or not he made a mean steak, how he liked his eggs. The literature does not show his relationship with his father, or if he was lonely. I imagine he was lonely. The literature does show that his coworkers left his side during lightning storms, afraid they would get hit. "See ya later, Roy," edging toward safety. The literature does show that Roy's wife was once struck beside him, as they draped the laundry on the line.

What does it mean to be stalked by electricity? To believe at any moment, a blast of charge may seek you out? I see him huddled. I see him eyeing the clear sky. (In a coffee shop, I look around. No one knows what I contain.) I print off Sullivan's astrological chart, repost it on the internet, but can find no one to read it. I stare at the symbols. I care because I want to know if it was something *in* him, if it is in me. Was he the bristling storm? Or was he just a man?

On the seventh bolt, lightning hit the top of Sullivan's head while he was fishing, alone. When he turned to his car, a bear tried to steal the trout off the line. Sullivan beat the bear with a stick.

Shock to the Heart.

When I said I felt clear on the soccer field, here is what I meant: I knew every cell in my body. I knew the cells made out of metal and the cells made out of protein. I knew the burned cells and the cells that would heal.

I knew it didn't matter what my settings were. I would die. We all would die.

I was no longer afraid.

Lightning Flowers.

In these three years, my whole body has regenerated: cells, skin, hair. So the machine, over the years, becomes more me than me. In an x-ray, its edges are clear, surrounded by the vague scribbles of my insides. It contains a copper stamp visible by x-ray so that, if I lose my ID cards, the doctors will know I belong to St. Jude.

And yet more and more, I know I belong to no one. In these three years, no doctor has felt the way this heart hitches when I speak to a man. No doctor has crouched with me on the office floor at work when my chest is quivering and thumping, when I am afraid I will fall. Though doctors have held my red heart in their latexed hands, there is no doctor who can teach me how to live. I was down on my knees in the soccer field grass. Only I could remind myself to breathe.

SAMANTHA SIMPSON

Blood Type

Blood Drive

MY BLOOD IS unexpectedly dark—a rich, almost-brown red. I gave away that first pint at the Moose Lodge in Mt. Vernon, Ohio, on an unusually warm morning in June 2008. None of the volunteers asked what had brought me so far from my home in North Carolina. They chattered with old friends and replenished the cookie supply while I watched my blood push through the thin needle burrowing into the swollen blue vein on the inside of my elbow. It charged through the clear plastic tubes, but it didn't have to travel far; it dripped into a pair of plastic sacks dangling from the lawn chair where I reclined. I squeezed a disembodied bicycle handle to keep all that dark blood flowing.

I had been warned: giving blood could make me light-headed, dizzy, or nauseated. A bruise could form at the needle site.

The skin of the volunteer nurse's hands reminded me of paper bags. I flinched the first time her cool, dry fingers traced the paths of veins on my forearm. I got used to her, though, to the way she peered at those tubes—full of my blood—and absently asked, "Are you okay?" She told me I didn't have to look at the needle or at all that blood. "Some people get sick," she said.

I didn't know my blood type. I'd left that line on the form blank—but that omission didn't matter. All of the American Red Cross literature told me that my contribution could save up to three lives. Across the room, women with gray curls and thick waists spooned meat stews into Styrofoam bowls and offered them to men with electric blue and lime green bandages wrapped tight around their elbows. Almost all of the people shuffling between the wood-paneled walls were white with deep wrinkles intersecting on their cheeks. They wore polo shirts marred by sweat stains, despite the low rumble of the air conditioner.

"Almost there," the volunteer nurse said. "Thank you so much for coming in today." She let me choose the color of my bandage, and I picked hot pink. "You can go over there and get something to eat, hon," she said. The flesh on my arm bulged around the bandage. She watched me sit up, then stand. She approved of the way I walked to the refreshment table without stumbling. She motioned toward the tables laden with treats. "Just take whatever you want," another volunteer said. "We sure do appreciate you." I picked through the pile of cookie packets—Oreos and Lorna Doones and Fig Newtons. I didn't speak to any of the other donors. I was apart—young and dark-skinned, a visitor from the college one town over. Nonetheless, the volunteer packed away my blood, and it would travel elsewhere to sail through the veins of an Ohio stranger.

Meanwhile, my mother—also a visitor and a different kind of stranger—made a list of family and friends to call from her hotel room in southern Georgia. The red digits on the bedside clock counted down the minutes until she needed to pay or check out. The money in her wallet had run out days ago—and my name was at the top of her list.

White Blood Cells

My mother is sick. She has been sick so long and so constantly that I can't think of a time when I couldn't write about her illness in present tense. I wrote about her brain surgery in purple crayon; I inscribed the story of her hysterectomy in a notebook decorated with cutouts of Brad Pitt and the cute ones from Boyz II Men. I wrote about her second and third rounds with breast cancer in leather-bound journals.

She has been sick for so long that I wonder how much longer I will be able to write about her in present tense.

Her sickness did not begin with the stubborn clump of malignant cells that blossomed in her brain in 1986. However, unlike her appendix, which ruptured soon after my mother's fourteenth birthday, her brain could not be removed. One day, she was teaching biology classes at a high school in North Carolina; then, in the nightmare days that followed, the names of her students slipped easily from her memory. The sound of her family's voices became cruel echoes that pressed against her skull. I don't remember if I was there when my mother learned that a thing in her brain could destroy her; I may have been crouching beneath the safe space of her chair while the doctor spoke.

She was thirty years old, and she chose the surgery—though I imagine she considered the option of wasting away in front of her husband as payment for his infidelity.

When she awoke from the anesthetic haze, her doctor's voice emerged from a too-bright light to reveal to her that he couldn't remove the tumor. Even now, it remains lodged in her brain tissue, a dormant volcano threatening to erupt too close to her pituitary gland.

Since 1986, more malignant tumors have latched onto her breasts. Her womb has nurtured more tumors than babies. Carrying me and my sister proved difficult; the burden of our bodies made it nearly impossible for her to breathe during those last trimesters. The cysts did not give her as much trouble as we did. They never distorted her body, and she could have them cut out. She didn't have to raise them alone. And those malignant cells have clung to her in a way that her adult children will not. Now they thrive in her marrow.

Blood Drive

The second time I gave blood, the volunteer nurse warned me not to look. The needle hovered just above the light blue vein at my elbow. I didn't look away. I watched the needle slide beneath my skin. The nurse rapidly pressed a bit of gauze to my elbow, and then she disappeared into a clean, white hallway, her paper coat swishing behind her.

In the fall of 2008, I worked two days a week as a lecturer in the English department at a university in Greensboro, North Carolina. I lay in bed while my roommate scuttled to work with tea sloshing out of her mug. I spent those mornings rearranging stacks of student work and planning classes a week in advance. But on this particular morning, I found myself reclining in a bright yellow lawn chair, watching more of my blood drain into a couple of plastic sacks.

Moments before, I'd discovered that my deep red blood carried rich stores of iron. The tip of my middle finger still ached from the sharp jab of the miniature needle the volunteer nurse used to extract one bright red drop of blood. She trapped the drop in a slide, which she slipped into a red box that whirred and clunked before it flashed the number 14.0 on its screen. "That's good," the nurse told me. I'd passed a test, which meant I could move on to a private room where I could answer questions about my personal habits.

I passed that test, too, largely because, at twenty-six, I'd already chosen a quiet life. I felt healthy and well that day. I didn't take aspirin. I had avoided receiving blood transfusions and organ donations in the past year. I hadn't had sexual contact with anyone with HIV or AIDS, and I'd never found myself locked in a passionate embrace with a prostitute. I managed to slink through both college and graduate school without getting that tattoo of a crescent moon on my shoulder. I'd believed in all the after-school specials that told me about drugs. I had never roamed beyond the boundaries of the United States; I tended to follow the same stretch of I-85 between my place and my boyfriend's apartment in Michigan. And, thus far, I had not inherited my mother's cancer. A few sickle-shaped red blood cells wandered lonely through my veins, but the volunteer nurse said they didn't count as a blood disease.

I'd abandoned that first donation in Ohio, but this batch of iron-rich blood would stay close. A heart in North Carolina would pump it through a local stranger's body. I imagined walking past someone downtown and feeling a tug of recognition.

Platelets

For nine days in August 1998, my family huddled together in a room at the Mark Inn in southwest Atlanta. It remains difficult to explain how we came to live in a place where my younger sister was not allowed to play outside. My mother, after all, had a couple of master's degrees. My sister read above her grade level, and I had just spent six weeks in central Ohio at a summer workshop on utopias in literature. We didn't belong there. Yet we were there, and we could not escape. No buses or trains ventured that far south, and we couldn't afford to hire a cab to take us to north Atlanta, where we could be more efficiently homeless.

That side of Atlanta sizzled beneath a merciless sun in those final days of the summer. If any thermometers strained and stretched toward the hundred-degree-Fahrenheit mark, I made no mention of it in my journal. I wrote, instead, about gulping down *Twelfth Night* while my mother prayed across the hall with a pair of fleshy women wearing worn-out sundresses. The air conditioner grumbled from its place on the wall, and we had to keep the curtains closed against the hungry eyes of local drug addicts and police officers.

In the evenings, which may have been cooler but humid, we three—my mother, my little sister, and I—paced the same stretches of thin carpet. I remember the arrhythmic blinking of the little red light attached to the phone and the snarling in my belly. We had once lived in a rundown house in Jonesboro, Georgia, but my mother and sister had abandoned it while I spent the summer reading philosophies in paperback. I wrote about being an "intellectual junkie" in my journal while we starved on white bread and bologna sandwiches.

Maybe my sister and I had to turn down the volume on the television while my mother made phone calls. As the cash in her wallet dwindled—and the hotel manager pressed her for another night's rent—she threatened to move us all back to North Carolina, where emblems of her failed marriage and doomed career seemed to rise from the landscape. But then her resolve returned, and she requested "just a little help" from the pastors and old friends who could spare it.

And maybe I bristled when my mother hung up the phone and shook her head. Our room smelled faintly of urine because maybe my sister—chipper as she could be during the daytime—couldn't help being nervous in the small hours of the night and wetting the king-sized bed we shared. I spent line after line in that journal wondering how people could be so cruel to us; my ink echoed my mother's anger at the fact that these folks knew—they *knew*—we needed help, but they couldn't even be convinced to send us ten dollars. "*Ten* dollars," my mother said. "And they act like they don't have it."

We certainly did not have it. In fact, we had very little that we could keep from slipping out of our grasps. We made an art of losing. We lost cars and books, baby pictures, and tchotchkes. We lost our clothes if we turned our backs for too long at the Laundromat. In the journal entry dated August 19, 1998, I wrote about losing that grim hotel room once and for all. Police officers escorted us to a parking lot glistening with rain while our neighbors across the hall made off with the suitcases and trash bags we couldn't carry. I did not write the way my heart felt as if it would break through my rib cage, and I did not write about the shame that slipped into my bloodstream and pulsated in my veins.

Blood Drive

I looked up the Ann Arbor donor center on a search engine. White and red lines intersected each other on the green expanse of the map on the screen. My point of origin and my destination were peach-colored balloons sprouting up cheerfully from the jumble of unfamiliar street names. I tried committing the route to memory, but I still jotted down the turns I would have to take.

I only noticed the thick layer of snow blanketing the earth and all the cars in the parking lot after I threw open the door of my boyfriend's apartment. In North Carolina, we did not contend with snow. If it arrived, we closed the schools, hoarded bread, and locked our doors against the chill. Here, the Arbor Land clock declared it was nineteen degrees. I piled on two coats, a couple of scarves, and a pair of knit gloves that still allowed the cold air to seep between my fingers. I worried about the slick patches of ice crouching in readiness beneath that layer of snow, and I imagined my car floating, then capsizing, in the snow. *I could die,* I thought.

Still, I ventured out. I had to do it: on the final morning of 2008, I informed my boyfriend that he was a sociopath because it had never occurred to him to donate blood. All that anger was born from the gaping hole in his bedroom ceiling. Melting snow had worn away at that boundary between us and the coldest Michigan winter in at least five years. At night, we lay head to foot on the couch while something scratched behind the walls. "You're selfish," I told him. I envied the way he could sleep through the night. I said, "You could save three lives with a pint of your blood." I felt my own blood warming beneath my skin in the face of his indifference.

And that was before the power went out and we lived on battery power.

That New Year's Eve, my car crawled from one intersection to another. The wipers whipped away the snowflakes that drifted onto the windshield while I leaned forward and gripped the steering wheel. I held my breath for much of that short drive, and then I trudged through a patch of fresh snow to reach the doors of the donation center.

The volunteer nurses welcomed me as if I were a guest arriving late to a party. They helped me shrug off my coat, and one of them complimented the rundown sneakers I'd slipped over two pairs of thin socks that let my big toes break through. During the interview, one of them noticed my student ID and asked me about my major. I smirked: "Actually, I do the teaching." She said that I looked young, and in that cramped interview room with its clunky computer, the transformation began. By the time the next volunteer swabbed the inside of my elbow with iodine, I had the face and legs to walk runways as a model. And by the time she slipped the needle into the spongy, blue vein on my left arm, I could add generosity to the list of things that made me gorgeous. "Wow," I said. "Thanks."

"Oh, no—thank *you*."

My new glamour wore off by the time I made my way to the snack table. I crammed a packet of peanut butter crackers into my mouth and chased it with a can of vegetable juice. No power at my boyfriend's apartment meant that we couldn't trust that the food in the refrigerator had survived the night. So I ate Fig Newtons and drank a bottle of orange juice. I grew hungrier as I hunched over those empty cans and cookie wrappers. Crumbs collected around my mouth. I didn't want to meet the gazes of the volunteer nurses, so I read about the different blood types on a paper place mat. A blood drop with a face smiled up at me from the mat; it, too, said it was grateful for what I had done.

Red Blood Cells

One evening in the fall of 1998, the phone bleated in my dorm room in Carrollton, Georgia.

I didn't hesitate to answer it. On my side of the receiver, I studied biology and welcomed any distraction from the cloying scent of the textbook's pages. I liked the way the ringing phone added to the coziness of the space I shared with my roommate and new best friend. We each made our beds. I propped stuffed animals against my pillow. She'd hung a collage made of fantasy book covers above her bed, and I'd collected more than a handful of anime posters and quirky postcards to tape all over my side of the room. I appreciated the cheap rug spread across the cracked linoleum and the unsteady stack of plastic bowls atop the mini fridge. I found as many opportunities as I could to say, *Let's go home.*

On the other side of the receiver, my mother surveyed her surroundings—she always had to be careful in that part of Atlanta at that time of night. Her place—another motel—always seemed dark, despite the dull green intrusion of the streetlights buzzing above her head. She called me because she had no one else to call. The crackle of the phone connection hid the way her voice broke as she asked for my help. She said she didn't want to call, and she didn't want to bother me. But her time was running out. This hotel manager threatened to put her out if she didn't come up with another day's rent on her room. "If you have anything," she said, "I'd really appreciate it."

I did have something: over a thousand dollars in refund money from a scholarship, a child's fortune. The cost of room and board at the university had already been covered. On most days, I felt safe, afloat on a small cushion of cash in my very first bank account. I said, "I can help," and I charged to her rescue. I searched for someone who could give me a ride to the Western Union inside of the local Kroger. On the way, we stopped at the campus ATM, where I pulled a crisp stack of hundred-dollar bills from my savings account. During those couple of years my family spent wandering from one hotel room to another, I had a difficult time imagining there could be this much money in one place and in my hand. We spent those years scraping together pennies and crumpled dollar bills. Now I had a handful of money—an amount that would have seemed impossible only two months before.

And that money was the least I could do. I gave my mother credit for engineering my narrow escape from the series of cheap hotel rooms that had become home during the summer of 1998. The effort cost her: she spent all of her money to send me to Carrollton in a cab, which meant she and my sister spent one night in a Wal-Mart. The night after that saw my mother sitting in a jail cell on the strength of a bad check. My sister had to be shuttled to Greensboro, where my aunt watched in horror as she hid food "for later." I had emerged from the crucible of our terrible lives. I no longer suffered, and I had done nothing to earn the comfort and safety my mother and sister could not have. Now I *could* do something; I could pay. I urged my classmate to drive faster. His headlights sliced through the darkness leading to the Kroger parking lot.

By the time I returned to the dorm, I was breathless. I dialed my mother's phone number with trembling fingers. She gasped when I told her the amount of money that I'd sent. She had only asked for enough to see her through the night, but I had sent her two weeks of survival. She cried now; she couldn't find the words to describe her relief and her gratitude. "I'll never forget this," she said. "Thank you so much." She promised to pay me back when she had another child support check. I wanted to cry too. I told her not to worry about it. With the steady crackle of our connection loud in my ears, I decided that I would be the kind of person who would do anything for blood.

Sickles

It could happen this way—

"I'm not saying it will," my boyfriend says, "or that I'm worried." But it could happen this way: My entire body could betray me just paces from the finish line of my first marathon. I would feel that betrayal completely: the clenching of every muscle, the buckling at the knees, the pulsing beneath my skull, the tightening and then failure of my heart. This betrayal would ring true and familiar in the seconds before my body—utterly defeated—would crumble to the pavement, suddenly dead. It could happen this way, but I say, "I'm training. I'm fine."

From my father, I inherited sickle cell trait. This usually means nothing. My blood races along the same courses while my heart pumps out its rhythm. When I was in elementary school, I pretended to be sickly to avoid gym class. "I could die," I would moan. "I can't do anything that's too hard on my body." I read stories about little girls like this—tiny ladies full of grace. I tried to be so graceful and gentle that birds would alight on my fingers and sing to me. I even "fainted" once when my mother sent me outside to play on a warm summer day. It was easy and painless to faint—only a matter of closing my eyes and allowing my knees to fold.

My boyfriend says, "I'm not worried, but you should be careful." The concern about my blood lurks on the edges of our conversation, much like the handful of sharp, scythe-like red blood cells sailing through my veins. They are few, but they can be lethal if they gather to block a blood vessel or attack my organs.

The thing about sudden death, he says, is that there's no arming against it. He says a body can just break down. I say, "I am careful." I don't need him to remind me of a body's treachery, especially when I know the damage my blood can do.

Hematuria

My blood was unexpectedly bright—searing red against the toilet bowl.

In fall 2000, I carefully listed my coursework alongside my part-time jobs in a spiral-bound planner. I did no reading for classes on Friday or Sunday. On those afternoons, I had to play on a trampoline with a kindergartner, and in the evening, I had to don a black and white uniform and serve hors d'oeuvres to professors and trustees. On weekdays, I shelved books in the library and scheduled interviews for a documentary on the college's science department. I could find no place in my schedule for a doctor's appointment to investigate the sudden appearance of blood in my urine.

I also couldn't find the money to pay for such an appointment. While I could delight in the gentle pressure of my academic responsibilities, I staggered beneath the weight of a seven-hundred-dollar debt on a Visa card. A little more than a year before, the credit card had arrived among a stack of bills in my mother's post office box. Although my full name was embossed on the front of the card, I'd hardly understood interest rates or credit scores. With the first swipe of that credit card at the airport where we rented the car that would carry me to college in Ohio, a new version of me was born. This version of me owed, and she continued to owe, despite the three part-time jobs I balanced in my daily planner. Each month, the minimum payment depleted the dregs of my bank account.

The debt did not shrivel, and the blood did not fade to brown or pink. I felt no pain in my lower back or anywhere else. I tried sleeping as much as I could, and I tried drinking more water and more cranberry juice. Still, all that blood escaped from my body. I attempted to calculate how long it would take me to bleed to death. Months, I decided. After all, I couldn't be passing pure blood—plasma, red blood cells, white blood cells, and platelets. I could keep working and keep studying.

In November, I cobbled together enough of my wages to make one substantial payment on that credit card. "Substantial" didn't mean much. Because I no longer cashed refund checks like the ones I received in Carrollton, I had to find ways to make a couple hundred dollars stretch for a semester. As I wrote the check to the credit card company, I tried not to think of the shoes and clothes I could buy—or the books and day trips to Columbus. I tried even harder not to think of the relentless crimson of the toilet bowl after every trip the restroom. The check shuttled to Delaware.

I waited for the next billing statement with a kind of grim delight. That one financial sacrifice left me hard and clean—or, at least, I felt that way. While my classmates anticipated care packages and checks from their wealthy parents, I took control of my own name. I fantasized about the version of me who did not owe, who only spent cash and kept her credit card tucked away for emergencies only. I imagined that next bill as a beginning and a confirmation, and when it finally arrived, I tore into it with the same excitement I reserved for birthday and Christmas gifts.

The new balance: over seven hundred dollars. And this was no mistake. My payment had posted, but so had a series of charges from a grocery store in Stone Mountain, Georgia, and the hotel where my mother and sister now lived. I kept my voice quiet and calm when I called home. I said, "What happened?"

A miracle, according to my mother. She and my sister had run out of food and out of time, and they didn't know what they were going to do. "So I decided to walk by faith," my mother said. "Walking by faith" meant walking into the grocery store and plucking cans and boxes from the shelves. "Walking by faith" meant pushing the cart to the checkout lane and testing every piece of plastic in her wallet. And the reward for "walking by faith" was discovering that one of those cards worked, that there was enough to cover the cost of groceries and the cost of another night at the hotel. She thanked God.

I said, "But I worked really hard."

A correction then: God used me to help my family, even if I didn't know that was what He was doing. For once, my mother did not sound exhausted and irritated. For once, there had been money in place and on time, and she could not squash her joy, even in the face of my disappointment.

There couldn't have been more than fifty dollars in my bank account on the day I scheduled an appointment with the specialist who would x-ray my kidneys for scars and damage. During the exam, I lay perfectly still while bright lights flashed above my body. I closed my eyes and, without protest, allowed a chill to creep under my skin. The procedure, which illuminated a kidney infection, cost five hundred dollars; this did not include the prescription that turned my urine pink then yellow then clear.

I shelved. I babysat. I served. I organized. And I owed.

Anemia

At twenty-two, I still stammered and blushed helplessly in the presence of all men. I couldn't help myself. I fretted over their nearness. The threat of banter of any kind—witty or dumb, sharp or limp—caused my throat to dry. I worried that something on my body—the curl of my lip, a stray hair, the flare of my nostrils, the nervous tears trembling in my eyes—would betray my unspoken desire to have someone's hand close over mine.

I was nothing short of hopeless at parties and bars, and my nervous babbling only grew worse when I found myself answering trick questions from bill collectors.

They always found me—or, they always caught up with the more daring version of me, who'd opened a Mastercard account in 2000 and went on a spending spree that only slowed when she hit her limit. She'd then flitted to another credit card. Her purchases didn't amount to much. She rented a post office box. She bought children's shoes. She charged biweekly payments on motel rooms. She bought groceries—baked beans and canned meat that simmered on a hot plate—and office supplies for the counseling service she intended to open in Decatur. "If you aren't the one who made those charges," one bill collector said, "then why haven't you reported your card stolen?"

I couldn't help fumbling a question like that, so I usually snapped my phone shut and scrambled back to my tiny and mean life in Greensboro, North Carolina—my hourly wages, my studio apartment, and my daily desperate search for a parking spot on a crowded university campus.

Having shed my name for her own, my mother rented a sweet little house in Stone Mountain, Georgia. My sister could zip along the quiet street on her bike, and my mother could wave at her neighbors while they planted bulbs or walked their dogs. She did not mention the thousands of dollars she promised I would pay with a hasty scrawl of her signature. Instead, she dreamed into the receiver. She intended to open a counseling service for people with drug and alcohol addictions. She intended to make millions of dollars and sit with Oprah Winfrey. She intended to pay me back—double, no triple—once she'd made her own way.

She dreamed while I watched my own ambitions shrivel. Despite my spot in an MFA program, I hesitated to call myself a writer. I worked as a substitute teacher, making sure not to go to the same school two days in a row because I could only afford one dress shirt and one pair of black pants. And when the bleeding returned after a four-year absence, I found myself openly weeping in the university clinic when the nurse said I would have to visit a specialist to uncover its cause. She estimated the cost of the visit and tests would come to about five hundred dollars. "Maybe seven hundred," she said. She might as well have said I needed a million dollars. Every credit card in my wallet had reached its limit. While my mother dreamed, I wondered yet again how long it would take me to bleed to death. She dreamed while my future became a parade of unpaid medical bills.

I only occasionally mentioned the collection notices landing in my mailbox. My voice grew small and shy, and I had to clear my throat. My resolve tended to founder in the face of her easy rage. "I told you I would pay you back," she said. She spent her evenings teaching DUI classes to a group of men and women serving mandatory penitence. She just needed to work hard and wait for her blessing.

"It's not just about the money," I said. "It's my credit." Before I moved to Greensboro, I had only a slippery grasp of credit reports and FICO scores. I believed in cash. While my gallery of collection notices expanded, my credit score dropped to 540. That pathetic combination of digits was proof that I was a risk. For most of my life, though, I understood "bad credit" as something that trailed my family, even though we were largely decent people. I thought "bad credit" meant unwarranted persecution. When my mother talked about her own "bad credit," she mentioned devilish bill collectors. "Bad credit," for her, was proof that she was not of this world—and no Christian was supposed to be, really. "Bad credit" was proof the world was against her after all.

But she told me not to worry about my credit. "I told you," she said, anger lacing her voice.

I found no room in our conversations to mention the dusky red clouds in the toilet. She and I never once talked about my doctors' visits or the cost of x-raying my kidneys. In one of the specialists' offices, I thumbed through a travel magazine with harrowing details of some mountain climber's journey. He clawed and grasped his way toward that peak—and hematuria became part of his trial. The higher he ascended, the more his body struggled against the menacing altitude sickness. None-theless, he reached such great heights. It didn't occur to me to mention the article to my mother either. On the phone, she dreamed and dreamed while I counted out pennies to pay the medical bills.

Blood Drive

"Uh-oh."

Only a moment before, the volunteer nurse had been all smiles and wheezing laughter. He told me about his children while he pricked my middle finger then coaxed a couple of re-luctant red drops from it. The machine whirred, and then his face fell.

"What is it?" I asked. "What's my . . . um, iron level—or, number?"

He shook his head, the fluorescent lights bouncing off the skin stretched over his skull. His face broke into a new grin. "I have an idea." He left, then returned with a packet of soft pads that heated my fingertips. "Use these," he said, and then he left once more, closing the door softly behind him. I squeezed the little pads, and then I sat on my hands.

The 2009 spring semester only seemed colder because the

chancellor delayed or canceled classes at the first hint of an icy road. No thick layer of snow surprised us in the mornings. Still, I had a hard time shaking the chill that had crept under my skin. I blamed the sliding door that separated my bedroom from the balcony in the new apartment I shared with my roommate. I piled more blankets onto my bed and wore another layer beneath my coat.

A new volunteer nurse swept into the cramped space of the interview room. She did not laugh, and she offered only noncommittal grunts to the small talk I made while she wiped the middle finger on my left hand with an alcohol pad. She cleared her throat before she pierced the skin and fed a new drop of my blood to the machine. It whirred—crunched—and then the number flashed onto its screen: 11.7.

"That's not going to work," the volunteer nurse said.

I pressed my thumbs to the bandage circles covering the tiny holes in my middle fingers. "What's wrong?"

"Your iron levels are low." She gave a little sigh. Her day had most likely been a long one. "You need to have at least a 12.5," she explained. "This doesn't mean that you're anemic. We just need . . . more." She handed me paperwork that advised me to eat more red meat and spinach.

All of that unexpectedly dark, almost-brown blood continued to charge weakly through my veins. I imagined shriveled red blood cells swirling past the sickle cells lingering here and there. I imagined strange particles surrounding and defeating my white blood cells. By the time I reached the parking lot of the blood donation center, I'd convinced myself that my chilled blood could hardly sustain me. I wasn't anemic, but I could offer nothing to the people smiling in the American Red Cross brochures. I imagined a shallow pool of blood drying in their hearts. I imagined their bodies fading into husks in their hospital beds.

This did not feel exactly like losing—and this shame did not feel exactly like the shame that followed me from that hotel room in southwest Atlanta to the dorm in Carrollton. The usual shame surprised me in those quiet moments at home, when the light of a single lamp illuminated a sheaf of frayed and bitter memories: Once, there were no Christmas presents. Once, we three—my mother, my sister, and me—vomited into the bushes outside of the moldy house that poisoned us all at the same time. Once, a tow truck carted away my mother's Skylark in the dark part of the morning.

This shame was different—but not unfamiliar. I knew it from the unopened collection notices in the mailbox. I knew it from my mother's desperate requests for money. I held my ground against my mother, but the bills flustered me. My bank account had run dry, and nothing could be done about it. Nonetheless, I still felt the nauseating charge of failure when I couldn't pay, and I felt it once more as I made my way back from the blood donation center.

O-Negative

It happened like this.

"Don't get upset." From there, the voicemail from my mother explained the phone call she had gotten from yet another debt collector. The story disintegrated from there. I sat in my tiny office on the university campus and squinted at the words I recognized: "Mastercard," "two thousand," and "lien." I hastily looked up that last word on the desktop computer in front of me.

And then I did get upset.

The shimmer of summer wore off almost immediately after I printed the fall 2008 class rosters. All those student names marched in pairs down the left sides of the pages. I dusted off the usual syllabi. I arranged and rearranged the particulars on my cv, hoping that in the coming spring, a new job would emerge somehow from the rubble of what would become the economic collapse.

Summer might have been fading, but I felt light, as if I could float on air. For once, I could believe that my happiness didn't cost anything. I didn't need to pay some cosmic price to nestle into the cat-scratched couch with my roommate or to blush when my new boyfriend's name appeared on the screen of my cell phone. Settling into that little apartment—which overlooked a clump of friendly trees that shaded a grocery store and a Tuesday Morning—had seemed easy; the adventure of signing the new lease and changing my address had been no adventure at all. I traveled between work and home, and, for once, no guilt nibbled at the edge of my mind. I knew I could deserve this little bit, this tiny corner of a good life.

And then my mother's ominous voicemail interrupted all of it.

"Your uncle did this to you," my mother said. Her voice became a deep snarl. "He *told* on you."

I gripped the phone and seethed. My roommate and I had already dimmed the lights in the new apartment. It was late. The mint green glow of the streetlights pierced the blinds. I paced in the near darkness, my heart pumping.

Earlier, when there had been daylight and clarity, I'd spoken to one of "the people" my mother had mentioned in her voicemail. They offered a hard deal: I could settle the debts that old version of me had incurred when she lived in a motel and charged canned goods to feed her eleven-year-old daughter. Or, I could have a lien on my car. The representative had been friendly and understanding—but firm. He said he wanted what I wanted: to end all of this, to make it stop.

Now my mother blamed her brother—an insurance salesman and minister with whom I only spoke when I found myself cornered and alone at family reunions. "He told on you," she insisted. "He's in on it. He's always wanted to destroy me." Even then, her battle with invisible principalities had begun to wear her out. She whittled down her list of friends, discarding women who had thrown rice at her wedding, giggled with her in college, lent her money, and reached inside of her to guide her daughter into the world. All of them had plotted against her, and now she had only her children left. She told me that these calls from the collection agency were a test and a plot. "They," she said, "are trying to get to me through you." This plot had hatched nearly twenty years ago, and her persecutors—who rarely had names—wanted nothing more than to stop her from changing the world for the better.

"No," I said. My blood warmed beneath my skin. "No, that doesn't make any sense. Why would Uncle Jerry want to put a lien on my car? Why?"

I imagined her gaping from her perch on her queen-sized bed. The days of shabby hotel rooms were behind her, and she lived comfortably in a rental house on a quiet street. Throughout the summer of 2008, she called me to ask for enough money to keep that house cool during the day and pulsing with warm light in the evenings. When I hesitated, she reminded me of that long-ago drive to Carrollton that depleted every cent she had. And I said, "Yes." She asked me for gas money, and she asked me to cover the cost of her stay at a hotel during a conference. I said, "Yes." I said yes even when my bank account buckled beneath the burden of my own obligations. Now I said, "No."

"Your uncle did tell on you—"

My free hand curled into a fist. "He can't tell what he doesn't know. And he wasn't the one who ran up that debt. I didn't even *touch* that credit card. It was you. You are the one who put me in this situation. And now I have to clean up your mess, and you're not even going to help me pay it. You're not even going to apologize." I tried to keep my voice quiet, but my near-whispers became something close to shouting. And I couldn't stop. By then, my mother had stopped gaping and begun her own shouting into the phone. I told her she didn't care about my future, and she told me I was ungrateful. I told her that I would never give her another dime—not ever—and she said I loved my credit score more than my own flesh and blood. One of us hung up on the other, and our shared history—the near-starving in southwest Atlanta, the tug of need, the relief, and the gratitude—disappeared into that night.

I paced the floor in silence, wishing I could keep hold of the summer sweetness that now slipped from my grasp. Now I noticed the unruly piles of unopened junk mail and bills on the dining room table. I would add to it a stack of papers to grade, and then the next round of credit card statements would arrive. Although only days before I had cleaned away a two-thousand-dollar debt on one credit card, I now had to begin again; I would cover for that distant version of me and spend a year making payments against yet another formidable debt. I said yes one more time.

I stopped pacing and dropped into one of the wobbly dining room chairs. I sat and allowed the pressure on my skull to build. I couldn't cry. This frustration was too familiar. I started ripping open the envelopes on the table. Some had been waiting for nearly two months while I spent the summer traveling between Ohio and Michigan for work and new love. I ripped old bills in half and set aside the ones that still needed to be paid. I found the occasional postcard, a couple of credit card offers—and then an envelope from the American Red Cross.

That cheerful June morning returned easily—the light breeze, the unassuming rays of sunlight filtering through the windshield as we made our way to the Moose Lodge. That morning, I'd said yes to giving away a pint of my blood and felt the tingle of anticipation of a new experience. Before that first interview, I filled out a form. I knew my own name and my address, but I hesitated in front of the question about my blood type. *What if I don't know my blood type?* I asked. I worried that this ignorance would stop the entire adventure in its tracks, but the volunteer nurse gave me a warm smile. She told me the need for any and all types of blood did not stop. *And you'll know your blood type when we send your donor card.*

Now that card had arrived on a tide of junk mail and bills. The card was attached to a letter with glue that yielded when I pulled at the plastic. The letter gushed gratitude and included a black and white image of a smiling little girl. Beneath my full name, I found my blood type: O-negative. I turned the card between my fingers and thought of my mother, who had once been a poet, whose body consistently betrayed her with its se-cret poisons—first, a burst appendix and then one cancer and another. I thought of my father and the skin that stretched taut over the heavy muscles of his thighs as he played aggressive rounds of tennis in Herman Park. In an old photo, he grinned in his uniform. His body had been solid, but his blood had been a haven for sickle cells. Those two had huddled together; their blood mixed and mingled. They created me—and now I stood close to alone, one of the handful of people in the country who cannot accept blood from just anyone. In an emergency, I could languish in the absence of O-negative blood.

Yet, I could marvel at the way those deep red, almost brown pints I donated earlier this summer now belonged beneath the skin of strangers. I stood close to alone, one of the handful of people in the country who can donate blood to anyone. My blood is unexpectedly accommodating. It can rush through any person's veins, nourishing organs and whisking away the waste. My blood—the blood that dies and renews itself, that returns and departs from my heart—must say, *Yes*.

MATTHEW FERRENCE

Mos Teutonicus

General Objective:
To provide the students with an opportunity to study the skeletons of
numerous animals they find in the ecosystem.

Background Information:
It is quite common for biologists studying wildlife populations to ex-
amine skeletal fragments to determine aspects of a species' natural
history. . . . In order to learn more about an animal from which a bone
originated, it is helpful to know where in the animal's body the bone
was located. Many similarities in bone structure exist among species
that have a common evolutionary history. . . . Anatomical structures
that have a similar evolutionary origin are commonly referred to as
homologous structures *by scientists. Once the observer has learned*
how the bones of a human are shaped and articulated, this knowledge
can be applied to other species as varied as a Blue Whale weighing 150
tons and a Pygmy Shrew weighing one-ninth of an ounce.

Procedure:
You will be provided with a disarticulated skeleton of an animal. Be
very careful to keep the bones of your set together. Make every effort
to avoid mixing bones from different skeletons or losing bones from
the set. Human skeletal models have been placed at your work area
for comparison study and to assist you in answering the questions
that follow.

(Gary M. Ferrence, "Skeleton Fragments,"
Fundamentals of Environmental Biology, 3rd edition)

I.

In the deep brush beside the stream, we stumbled across bones. Hidden from the farmhouse by the steep hillside, my friends and I could easily have imagined ourselves lost in the deep woods, kicking away dried grass and carefully pulling aside spiked thistle until the intact skeleton of a deer emerged. Leathered hide clung to the bones, tatters of dried skin and bristled hair. The smell wasn't bad, just the faint sweetness of long-rotted flesh. So we wrapped our fingers around the antlers and pulled. The last fingers of grass held for a moment, then tore. We dragged the deer out of the stream valley straight up to the house. We felt this would be a possession worth the effort, even if its value was not yet clear.

This was the summer after ninth grade, the year my chubbiness melted and my shortness grew away. I'd stretched to five feet nine inches, no monster, but tall enough to imagine a towering future. I weighed only 120 pounds, so my own bones poked at my hide, particularly the knobs of my wrists. There, I could easily overlap my index finger and thumb, pinching across the narrow space and wondering when it might thicken into the beefy farmer's muscles of my father. I didn't know that I'd reached my full height, nor that I'd eventually swap the anxiety of the scrawny for the worry of middle-aged paunch.

Dragging the deer skeleton toward my house seemed like the right kind of challenge for a group of kids infatuated with *Dead Poets Society*, a film that less than a year before had encouraged our vaguely literary aims. The deer carcass seemed like something existential, or at least a sort of limited proto-existentialism accessible to nascent high schoolers. Its grittiness offered a morbid and alluring invocation of mortality. We all came from upper-middle-class families that created children drawn to such notions of existentialism, even as our lives were bereft of the suffering that goes along with it. Of the seven in the group, six had at least one parent who was a professor at the local university, and the other had a chiropractor father and a mother who happened to be the AP English teacher at the high school. Even my father, the farmer, was really a biology professor by day.

When we first found the deer, it was my friends who had wanted to claim it. I'd wondered why, even balked inwardly at the rightness of the act. Partly, the deer was just another carcass to me. My father had killed many over the years, driving a neat hole through their lungs with a .30-06. A dead deer hung in the backyard walnut tree each winter, its smooth muscle exposed to the air after my father yanked the skin and hide free. He hacked it first into rough quarters with a sharp hatchet and a large knife, then he and my mother used smaller knives inside to divide the hunks into cuts of meat headed for our freezer. Our dogs gnawed the leftover bones for weeks. My brother and I had each shot deer too, as well as squirrels, and rabbits, and pheasant, and groundhog, and sometimes throngs of gathering autumn blackbirds that fell in clumps after each shotgun blast.

None of my friends had likely ever shot a gun, nor even touched one, save John. Sometime during elementary school, I'd snuck him into my parents' bedroom, withdrew a leather gun case from beneath their bed, carefully unzipped it, and showed him the smooth walnut stock of one of my father's rifles. There was magic in that moment, I thought, the revelation of the dangerous, the violation of the house rule to never touch a gun without my father present, the sweet tonic of gun oil and the burl of polished wood.

My friends did not know the feel of a dead animal, had never killed, nor slaughtered. The dead deer drew them in, made them adamant. In particular, a girl for whom I still harbored desperate adolescent love wanted that deer, and so I wanted it too. My friends sensed their own magic in the unfamiliar proximity to decay. It was this, I believe, they wished to possess, maybe even contain.

We dragged the animal to the house, deposited it by the back door. My father came outside, regarded the skeleton with the combined pride of hunter and biologist. *That's a big one,* he said, *and it's all there.* Immediately, he noticed the tatters of flesh clinging to the bones, smelled the unmistakable cloud of rot. He suggested that we find an ant hill and drag the deer atop it. The ants would emerge, he explained, to eat away the remaining flesh. In a few months, the bones would be picked clean, and we'd be left with a pristine, entire skeleton. Ants would work as a kind of natural process of *mos teutonicus,* the medieval practice of boiling away royal flesh to safeguard remains. Long ago, bodies in need of transport had to be reduced to bone to prevent rot, as the appearance of decay was a visual reminder of the wickedness of impermanence, and, I imagine, more than a little unpleasant to travel beside. First boiled, then flayed, the bone became free of flesh, of putrefaction, of the reminder of ever-present mortality. Afterward, the skeleton could be carried long distances for proper burial without fear of degeneration.

We followed my father's directions and dragged the deer across the driveway to the old orchard. We found an ant hill a few yards away from the compost heap and centered the deer across the opening. We would return. We would check it often. We would experience the enlightenment of excarnation.

II.

I was six years old the first time I broke an arm. We hadn't moved out to the farm yet, but I spent plenty of time there while my parents worked to make the house more or less livable. My friend Jeff came with me one day, and the two of us crawled around the barn, eventually winding up hanging on the front paddock gate. We began to play there, letting go, falling back, catching ourselves with our fingertips. I remember the giddiness of free fall, a slight twist in the stomach as my back rushed toward the ground followed by the solidity of my fingers catching wood.

I don't quite remember falling, nor precisely how it happened. For more than thirty years, Jeff and I have joked about the fall as a consequence of his deciding to up the ante and shake the fence. I'm not sure that he ever actually did that, and some corner of my memory suggests that I invented the story to prevent my parents from punishing my carelessness. Whether he did or didn't shake the fence, I failed to catch it and, instead, flung myself upon the concrete threshold between barn and yard. While I cannot remember the feeling of impact, I can easily recall what followed: a sharp pain, then the dull sensation of a heavy, useless limb, leaden, an arm suddenly external from my consciousness.

III.

To x-ray a one-year-old, a technician must first promise a sticker from a ten-cent sheet of low-grade *Toy Story* images, maybe two if he feels particularly generous or guilty. To earn that, the crying one-year-old must lie down on a dingy metal slab, then submit to having an industrial-gray strap tightened around his hips. The one-year-old will be instructed to lie still, and the technician will scowl at the involuntary heaving of shoulders that might spoil the shot. The father must don a heavy lead apron to protect his own testicles from piercing radiation, all while considering the destruction of his sperm reasonable: no worthy parent should stand by and subject his son to this sort of procedure. There will be a mechanical buzz, audible because the child's sobs have turned to wide-eyed fright. The sticker will be received, grasped in the boy's left hand and regarded with moist eyes. His smile will seem like an indictment, an unearned forgiveness offered too easily.

Next, a rushed pediatric orthopedist will remark on the oddity of seeing cloth diapers in Western Pennsylvania. *Are you from Seattle?* he will ask the parents. The father will recognize this as an insult to his home region, an implication of the backwardness of Appalachia in comparison to elsewhere, but he will also feel pride. He and the mother *are* Seattle kinds of parents, which will let them ignore the orthopedist's condescension or, perhaps, share in it. They will pretend to be breezy and calm during this brief small talk that further delays the information they have come to learn. Already, they will have driven an hour and a half to a suburban satellite of the city Children's Hospital, and they will have brought their happy one-year-old into the bland business park, and they will have watched him grow steadily bored and cranky while he waited an hour past the appointment time for which the parents have been instructed to under no circumstances be late.

With the orthopedist, the father will speak in a version of overly precise medical jargon, explaining how his son *presents weakness* in his right hand, that he only ever crawled commando-style, dragging his right arm and leg while he chased the dog around the house. The parents will explain that their child has merely expressed an early preference for left-handedness. The father will attempt to demonstrate that, while he is not from Seattle, he is in fact from rural Western Pennsylvania, he is intelligent, that he has a doctorate too, albeit as a specialist in literature instead of bones, terrified that not demonstrating such intelligence will lead to the disdain of the doctor, which was probably inevitable anyway. The mother will explain how a family friend, a doctor herself, noticed the way the boy's right foot dragged when he walked and suggested this appointment. The mother will confess that her own father had noticed the boy not using his right hand as much as his left, and that the mother and father discarded the information as another of the regular paranoias of a chronic worrier.

The orthopedist will ask the boy to walk down the hallway, and he will watch as the boy complies. The one-year-old will be happy again. He will walk and grin, turn and walk back. He will turn his right foot forty-five degrees, and the toes of that foot will drag along the industrial carpet. The orthopedist will frown in the way doctors frown. He will turn abruptly back to the examination room and sit. The boy will return also, into his mother's lap. The father will think about the lightbox on the wall, and about his own familiarity with orthopedists. He will recollect spinning cast saws, and the soggy odor of fresh plaster, and the ghostly vision of his own bones hung on the wall. He will remember slowness and fright and the sharp smell of antiseptic. The orthopedist will flick on the lightbox on this wall, and the ghostly vision of the one-year-old's pelvis will appear. The father will pretend to understand.

First of all, this isn't orthopedic, the doctor will say. He will gesture at the x-ray and explain how there are no signs of hip dysplasia, that the leg bones are normal. The structure is good. He will run the flat side of his reflex hammer along the underside of each of the boy's feet. He will nod to himself, confirming something he suspected when the left toes curl downward as the tool scrapes from the heel forward and the right toes curl up, making the Babinski response. The father will not know, yet, that Joseph Babinski was a late-nineteenth-century French neurologist of Polish ancestry. He will not consider the coincidences of convergence: of French for the mother's career and Polish for the father's heritage and Joseph as a mutual name for neurologist and child. The father and the mother will not understand, yet, why the orthopedist nods after the test, why he refers the child to a neurologist, what he means when he says, *It will become orthopedic. They always do.* The orthopedist will begin speaking about *hemiplegia.* This he will not explain either, perhaps convinced by the father's earlier feigning use of medical jargon. The father does not understand the term. Afterward, the parents will drive an hour and a half home, the weary boy falling asleep within minutes of being strapped down again, this time into his car seat.

Later, on the phone with another friend who happens to be a doctor, the father will scoff at what the orthopedist implied. He will say that it probably means nothing, and the friend will agree, until the father mentions Babinski. Then the friend can no longer be just a friend, and must now be a friend who is a doctor. *Oh,* he will say, *that's different.* The friend's tone of voice will change, will become a bit more formal, and his timbre will flatten to indicate seriousness. His voice will be that of a doctor now, a good and caring one, but precise. The friend will explain what the orthopedist did not or would not explain, that the Babinski response is a clear sign of neurological trauma, likely cerebral palsy, *hemiplegia* referring to the affliction of only one side of the body. A neurologist will later confirm this diagnosis.

To x-ray a one-year-old, the father must be prepared to learn what he does not wish to know.

IV.

One afternoon when I was in third grade, I climbed onto a fallen log that crossed the stream at the bottom of the hill below the farmhouse. I stuck my arms in the air, walked as if on a balance beam, then slipped on the mossy bark. Moments after landing on a stack of old boards rotting in the grass of the lower pasture, I raised my right arm into the air and felt the unmistakable heft of a break. I walked back to the house, tearless, and informed my mother that I'd broken my arm. She did not at first believe me, considering the calm in her nine-year-old. But she knew my arm hurt enough for a trip to the emergency room, where the doctors confirmed broken arm number two.

A few weeks later, still wearing the cast protecting the chip in my elbow, I chased a ball into the wood-mulched flower garden surrounding my friend's house. It had been raining, so the wood was slick, and my balance was compromised by haste and plaster. I slipped, could not catch myself since one arm was occupied by a ball and the other was out of service. I banged my head on the glass cover to their electric meter.

Dazed, I walked into the driveway, where John looked directly at my head and began to cry. When the first drips of blood began to fall across my face, I began to cry too. His mother was a nurse, and she dragged me into their bathroom and began pouring water onto the wound. Blood filled the sink as she assured me in her heavy Greek accent, *One stitch only, maybe two,* explaining that head wounds always bleed this way.

V.

Cerebral palsy, or CP, is a specific-sounding term for conditions that range widely in both manifestation and severity. The Latin roots of the term identify the underlying cause of such conditions: *cerebral,* as in dealing with the cerebrum portion of the brain, and *palsy,* paralysis. The expression of cerebral palsy, however, does not involve physical paralysis, or at least not technically. Instead, cerebral palsy is categorized into two main types, with further finer distinctions. *Spastic* or *pyramidal* cerebral palsy designates chronic muscle tightness, like a permanent muscle cramp, and results from damage to the strands of nerves running from the cortex to the brain stem called pyramidal tracts. *Non-spastic* or *extrapyramidal* cerebral palsy designates decreased muscle tone and is often associated with involuntary movement or tremors. In these kinds of CP, the damage has occurred outside of the pyramidal tracts, and is further subdivided into *ataxic* and *dyskinetic* cerebral palsy.

CP can affect both motor control and cognitive ability.

CP is non-progressive.

As an injury to the brain, often from a stroke, CP is without cure. Unlike a bone fracture, which can be set to ensure proper healing, a damaged brain does not regenerate.

In describing CP, the prefix *quadri* refers to impact on all four limbs—in fact, to both sides of the brain—while *hemi* refers to impact on two limbs, due to damage in only one hemisphere of the brain. The suffix *plegia* refers to paralysis, and *paresis* to weakness.

VI.

For my third broken bone, I merely fell out of bed in the middle of the night, smacking my right arm against the heat register. My father called home from Wyoming, in the first leg of a sabbatical tour of the western national parks. My mother filled him in on the news. *Your daughter bought a new puppy. Your youngest son broke his arm again.*

I was in sixth grade, and when I returned to school I listened, mortified, as the teacher explained to the class: "Matt got into a wrestling match with his bedsheets."

VII.

My friends never looked for the deer skeleton again. I checked on it occasionally, looking in on the ants' progress. My father had been right. By the end of the summer, the tattered flesh had been eaten away, leaving only sun-bleached bone. My friends never saw this, nor asked about the skeleton, nor apparently ever thought again about the deer after its discovery. There, too, we considered the deer differently. To them, it had been the find of the moment, the excitement of a singular occasion. It had been about the thrill of a deer skeleton. For me, who cared little for the specifics of deer, the thrill had been something else, a threshold of mutual discovery and connection and of future excursions into ourselves. The deer skeleton was about the group, about being part of a group. The farm had long thrust isolation upon me, reinforced by the usual sense of adolescent awkwardness. The group, which melted away just as late summer afternoons gave way to chilly fall evenings, had offered a welcome sense of community. These were my people.

We stayed friends, and in some ways close friends, but things were soon different. Shortly after we started tenth grade, the others found upperclassmen who practiced a clearer and more direct sort of alternative artiness. They were professors' kids, too, but ones whose greater age and clearer sense of self had more appeal. Those boys also had a band, and to this I had no defense. My friends started hanging out in town.

They included me once, long after the new relationships had been forged. We spent a Friday night driving dark streets in a rickety van. We snatched a shopping cart from the parking lot of a grocery store, feverishly pulling it through the rear doors of the van before speeding off. The older boys wanted to stash it in a weedy ditch, then return another day to take moody photos. This was an aesthetic sophistication with which I was neither familiar nor fluent. They laughed and marveled at what they would do, while I bounced on the floor of the van's cargo space.

I was bored, and it became clear to everyone that I wasn't really part of this new collection.

Though there came no moment to pinpoint the exile, I felt cast out. This exclusion felt inevitable to me, a consequence of growing up on the farm, where location posed a problem. My friends could walk to one another's houses, and they enjoyed spontaneous gatherings, street hockey and neighborhood games of mass tag, and trick-or-treating by foot instead of by car. I always needed permission from my parents to see friends, always needed a plan, and a schedule, and someone to drive me. This was an imposition, however unintentional, that created a certain childhood loneliness and an expectation of being left out, partly because I often was. I wandered the farm by myself, playing a game I called German 88s, carrying one of my arsenal of toy guns as I pretended to be a lone GI dropped off behind enemy lines to destroy Nazi positions.

In many ways, my feelings of high school exile were common. We all felt that way, I'm sure, even if I considered my geographic isolation as different and more profound. In the years that followed, the others were part of a new crew of friends who often visited the farm, particularly once we could drive. We played nighttime tag there, where I donned camo pants and a black T-shirt and distinguished myself as expert at hiding, never to be found. Eventually, such nights always wound to a sleepy close, my friends driving down the valley back home, while I watched the flash of taillights disappear around the bend of our dead-end street. I stripped off my dark clothing and felt the quiet of a farm night as equal parts solitude and loneliness.

The deer skeleton lingered in the weeds of the old orchard, long freed from remnant flesh. Night animals crawled among the same low branches where I had spent hours hiding, crickets and katydids and cicadas and frogs sounding the dark. On some nights, I listened to the ethereal chirps and screams of raccoons, or the lowness of a distant owl, or the rustling of unknown creatures searching the shrubs beside the house. Each seemed to reinforce a sense of separation that, even now, I find hard to embrace, that there is a difference between bone and flesh, that what lies at our center may make us ourselves, yet we live our lives mostly chasing tattered flesh. And high school remains forever the place where flesh is tattered. Eventually, the deer bones began to scatter, too, pulled apart by the raccoons and by coyotes, until nothing remained.

VIII.

Break number one was on the radial styloid process of my right arm, called a chauffeur's fracture. My age and the severity of the break meant that the arm had to be set under general anesthesia. I lay on the hospital bed, watched the black mask approach my face, smelled the rubber as it covered my mouth, then drew in fumes, a brief spinning in my head before, faster than is natural, I slipped under.

Break number two was on the radial head of my right arm, which in imperfectly healing helped me learn my right from my left. Faced with the question of turning one way or the other, I learned to check orientation by bending my elbows to reach each hand toward its respective shoulder. With my left, I could lay the palm flat. With my right, the arm stopped short, caught up by new bone, so I could barely brush the shoulder with my fingertips, a permanently limited range of motion.

Break number three was a hairline fracture of the left radius. At my first visit to the emergency room, the doctor guessed that my description correlated with a break even though the x-ray appeared fine. My prior experience of fractures, perhaps, convinced him to trust my intuition, so he applied a temporary cast. The next day, the orthopedist sawed off the cast to shoot another x-ray. Enough time had passed to allow the thin fracture to open and appear on the film as a bent worm. For the first time I received a fiberglass cast instead of a plaster one. While light enough to let me almost forget its presence, its gnarled surface also resisted signatures, the ink soaking into the dents and making my friends' practiced scripts ghostly.

IX.

Because our son was only a year and a half old and unlikely to hold still for the extended periods of time an MRI requires, a CT scan was the only option to make a sure diagnosis. This would likely have required sedation, the numbing of the senses so we ourselves could gain a sense of surety.

We balked at the tests, concerned by the risks of sedation and, worse, the potential cancer risk associated with the relatively high radiation dose of a CT scan rarely calibrated properly for a young body. The neurologist assured us of the prudence of that decision, that the tests were really only confirmation. Physical examinations had already told him enough to make a diagnosis of hemiparesis. Unless our son started to show more severe symptoms, the neurologist was confident in ruling out a tumor or other sort of progressive problem. He'd watch for change, but unless things got worse, there was little reason to worry.

Worry is, of course, not built upon reason. Though firm in our desire to protect our son from further damage or fright, we couldn't quite shake the fear that we weren't proceeding as we should.

Proceeding.

Procedure.

The precise language of medicine facing off against our growing recognition of the imprecision of life.

My wife and I came to realize that our decision was not one of proper procedure but, instead, of philosophy. Too often, disorders and diseases and conditions and all sorts of medical diagnoses appear technical. The body takes on the appearance of a machine, breakdowns attributed to worn or faulty parts that can be swapped or fixed. Physicians, then, function much like automobile mechanics. Diagnoses are made, followed by repair or replacement. The soul of the machine is a secondary consideration or, too often, considered irrelevant.

We chose instead, not without difficulty, to recognize our son's diagnosis as a description of his traits instead of an ailment. As our first appointment revealed, CP is not orthopedic, and I take that as a way of saying it is not structural. His bones are not affected, and there is nothing to fix. We choose to focus on a little boy. The memory of his first x-ray makes it clear enough that a child's soul is often overlooked in favor of medical diagnosis.

We could have strapped him down again, or allowed the flooding of sedative into his tiny veins, and bombarded his tissue with radioactivity. We could have looked at a colorful image of his brain, light spots showing function, some dark spot revealing where the damage had been done. But this would tell us little we didn't already know, and it would tell us nothing about him. We could have relied on the technical as a way to avoid considering the actual structure, focused on the flesh of diagnosis instead of the bone of a life that will be lived with challenges different from our own.

Our son has never doubted his ability to live as a boy, has never even considered himself as anything other than as a boy. This is the wisdom of the child, something we should all aspire to, and it is no folly. He is all child, a climber, and a thrower, and a laugher, and a joker, and so many things never visible via medical imaging.

And, really, I know enough about the origins of his CP. I will never forget the half-light of the delivery room, many hours into a labor that my wife faced, by choice, without medications. I will never forget the exhaustion on her face as our son's head kept appearing, then receding back inside the birth canal. I will never forget how I prayed mightily, how I asked my long-dead grandmother to pray as well, how I hoped that the connection I somehow shared with her across the life-death divide would help. I will never forget the look on the midwife's face when she showed me the blue-gray umbilical cord, knotted in the center. She'd never seen one pulled so tight, she said, and I understood the implication of the words unspoken: never seen one pulled so tight on a baby delivered alive.

A body is structure and soul. I think about the literal structure of our bone as the finest metaphor for our truest self. Bone is alive, hardened and apparently rigid, yet teeming with a marrow and blood. Bone is alive as a soul is alive.

I think, too, about the misdirection of flesh, and appearance, and the temporary satisfaction of something that appears meaningful. I think about the old deer skeleton, and the eventual rotting away of skin that accompanied the erosion of friendships, and my own inability to recognize the difference between surface and depth. I think about the surface knowledge of tests and how it compromises the desire within bone, seeks to boil away the flesh to see permanence. I think about how bone is not the end, that within it lies more life, cells and blood and soul. Tests seek the rigidity of the absolute, implying first that the absolute can be known and, further, that intervention is both necessary and helpful. Tests cannot find a soul, which merely desires the space to find itself.

Our son has been on the move since learning to walk, a desire that many have labeled as an instinct for self-therapy. His desire to run and jump has been recast from a boy's energy to therapeutic exercise, just as too many holiday presents from too many well-meaning family members have been offered with some eye toward their potential to improve his range of motion. But children do not exercise; they play. A ball is not for therapy, a board game not for right-hand-dexterity drills, a push wagon not for walking support, a scooter not to build leg strength. And a bicycle is not a tool to combat CP; it is a bike.

Our son does not "stretch his limits" or "commit to his own therapy." He does not think of clichés of birds being able to fly because they don't know they cannot, or the merits of facing obstacles and pushing through. Cliché is tattered flesh. Orthopedists are tattered flesh. CP is tattered flesh. CT scans are tattered flesh.

x.

Two decades after converting to my father's Lutheranism, my mother returned to the Catholic Church. There was a family meeting, and each parent explained the reasons for the theological split. The three children were given the choice of which way to go. My brother and sister remained Lutheran, and as a ninth grader driven by the clear appeal of homilies described to me as shorter than sermons, I became Catholic. Functionally, there was little difference between the procedures of the high Lutheran service we grew up in and the Catholic Mass, but even then I understood that, for my parents, the differences were keen and fundamental.

The timing of that conversion and the discovery of the deer skeleton seems to be more than coincidental, as entry into the Catholic Church also marked my entry into a theology that honors the tradition of reliquaries. Where else should saintly souls reside than in the gnarled remains of bone and sinew? As Brenda Miller writes in "Basha Leah," what bone scientists call *vertebra prominens* and Jewish faithfuls call the *luz* is an indestructible hub at the base of the spine from which our bodies will be rebuilt after death, and also a point that faces skyward when we bend forward reverently, when we humble ourselves to the unseen. How then could I ever deny the faith of French friends in Lorraine, who once proudly explained to us that their church possesses the knuckle of an apostle? How could I explain away the rush of emotion my wife felt when we lived in Paris, on the day she and my mother leaned forward to kiss the Crown of Thorns venerated at Notre Dame? Soul is bone. We are bone, forever ourselves, and permanent, and alive so long as we refuse to submit to the misguided allure of tattered flesh.

I am convinced, finally, of the presence of my Catholic grandmother's soul twice. First on the night she died, when I wandered into my parents' room, still asleep, and wordlessly awoke them a moment before the phone rang to inform them of her death; second in the delivery room, breathing life through our son's constricted umbilical cord, loosening the knot enough to usher him forward, alive and breathing and crying, as newborns should, ready to take immediately to his mother's nipple and draw in the first gulps of his first food, her soul feeding his, my soul open and grateful.

SARAH VIREN

Wolf Biter

When you hold out your hand and the whole world stops and
you find yourself / looking at the back of your hand, which,
the longer you look at it, looks starved.

—Cole Swensen

I AM IN a theater. Illuminated by stage light, the graying novelist leans into the podium as she reads, rocking the audience back in time, across the Midwest, to a little house in a nothing town where a family plays checkers. In one of the back rows, I lean into the plush chair, holding one hand to my side and out into the aisle as I peel back the loose skin once rooted at the bed of my thumbnail.

My mind wanders. I am rapt. My index and middle finger dig, pull, and tear at my thumb. The author flips the page and turns to the subject of home—what it means, how it morphs and eludes us—and I abruptly recall the world around me, the rows of seats behind me, the people everywhere who could be watching me. This, not pain, stills me, though not before teeth have torn flesh. By the end of the night, as hands meet each other in applause, my thumb gleams soft pink, a throbbing corona of what look like stretch marks around the nail's edge.

The first comprehensive analysis of the human hand came as a proof of God. Francis Henry Egerton, the Earl of Bridgewater, on his deathbed in 1829, commissioned a body of work meant to prove that the universe came with a celestial design. In life, the bachelor earl was a careless, jovial man known for hosting canine dress-up dinner parties and insisting on a different pair of shoes each day. But in death the earl wanted to leave a legacy of teleological proof of the Divine. One such proof was a tome devoted solely to the human hand: "The Fourth Bridgewater Treatises: On the Power, Wisdom and Goodness of God as Manifested in the Creation. The Hand: Its Mechanism and Vital Endowments, as Evincing Design." Its commissioned author, Sir Charles Bell, produced 428 pages of analysis, methodically leading the reader through the curve of the human shoulder to the tips of the fingers, all the while comparing the shape and utility of these parts to similar structures in elephants, sloths, camels, cows.

"[I]t is in the human hand," he wrote, "that we perceive the consummation of all perfection, as an instrument. This superiority consists in its combination of strength, with variety, extent, and rapidity of motion; in the power of the thumb, and the forms, relations, and sensibility of the fingers, which adapt it for holding, pulling, spinning, weaving, and constructing; properties which may be found separately in other animals, but are combined in the human hand."

Reading Bell's encomium to hands, you cannot help but pause every now and then and look down at your own. Those ten digits. These tiny tools. Hands are not only mechanically perfect, Bell argues, they also possess "the property of touch, by which [the hand] ministers to and improves every other sense." It is touch that does me in. If I only looked at my fingers, I would never bite or tear at them again. It is when I am not thinking but feeling, when my fingers sweep check one another for loose or uneven skin, that I feel compelled to trim. I want to fix the errors of my skin, and so you might say that I, too, believe in the perfection of the human hand. The only problem is, I don't know where to stop. In trimming, I inevitably leave a scrap of unevenness, some unruly skin or nail bit that then forces me to return, readdressing the issue again and again. And with this I am Sisyphus. My hands, my boulder.

I had a lover once who winced when she looked at my hands. We might be in the car, and she would glance over as I shifted from second to third. Or maybe making dinner in her cumin-scented kitchen, me chopping carrots and she stirring lentils, and she would lean over to share some endearment but get distracted by the injuries lacing my fingertips. In these moments, she covered my habit from her eyes, her hands encasing mine.

With her, I tried for the first time to stop. I bought Band-Aids, cycled through packs of gum. She gave me organic ointments, with which I rubbed my fingertips to a shine, and when spotting signs of healing, she complimented my hands with a mother's pride. For a while this worked. I stopped biting and tearing in her presence. And when we took a road trip one June, the three-week abstaining period allowed my fingers and fingertips to heal and my nails to grow thick and white-tipped. I chose to see this as a personal triumph, an evolution of self—until months later, that is, after I'd moved across the country and broken off the relationship and suddenly, almost without even noticing it, began attacking my hands again.

When I try to identify how all this began, I remember when I was seven or eight and my neighbor, a towheaded girl named Stevany, began to bite her nails. She nibbled at them with bravado for nearly a week before I finally asked what she was doing.

"Biting my nails."

"But why?"

"Because that's what people do."

Stevany was two years older than me. She and her even older sister Dee taught me to play Spades and told me my first sex joke. In the winter, they gave me their hand-me-down clothes, and in the spring, they showed me how to drink the beads of sweet from the honeysuckle. After Stevany explained nail biting, I spent days mulling over its potential, waiting until a moment when I was alone to try.

The initial bite came off, I realized later, much like a first cigarette, everything unnatural but exciting, as if my body had a voice and that voice proclaimed: "This is not right." But then a whisper urged more convincingly: "Do it again."

And just like that a habit formed. Then it progressed. Until one day I realized it was less habit and more disorder. *Dermatophagia* is the technical word: skin eater. Another term for people like me is *wolf biter*, although no one in the medical literature explains why. Are fingers like wolves? Does biting and tearing flesh make me wolf-like, make me wild?

The crux of it, for me at least, is this: I either see, or with my other fingers feel, a loose piece of skin dangling somewhere around the nail. I am then aware of this flap and yet for some time avoid it, avoid thinking about it, a challenge I meet with success until that moment when I fool myself into believing that I am acting subconsciously, and then I pull, I rip, I tear. Skin removed from flesh. And what do I accomplish? Order. A sense of serenity. Satisfaction.

It's not the act of tearing or biting I crave beforehand, nor is it the pain afterward. What I want is the satisfaction of a desire filled. The forcing of errant parts of myself into submission. The belief that I am the maker of my hands. This is textbook obsessive-compulsive behavior. I know this. And yet, if we're going to use labels, I prefer wolf biter.

Another lover bit her hands even worse than me. A jet mechanic in the air force, she worked graveyard shifts, would drive to my bedroom at dawn, her uniform baked to a metallic sheen, her fingers washed in grease that collected near nubs of nails where years of biting had inflamed the tips so much they splayed. After a shower, her body lowered into the bed beside me, still smelling slightly of hot coins, the creases of her hands shaded bluish gray. Waking, I would hold her fingers against my palm, mesmerized by the brutality.

"Why do you do this?"

"Why do you?"

By comparison, my hands gleamed porcelain, but she had a point. When our habits deform our bodies, we can't hide the proof of what we do. Still, I wanted a distinction, a hierarchy of disorder. Even today I can see her in my mind, the way she would chew at her fingers with a determined cock of her head, a slight clench of the jaw. Her eyes always looked out in the distance, as if meditating or remembering.

"Jenn," I scolded. "Stop."

And then her gnawing would cease for a few moments or hours, but at some point her fingers always returned to mouth—just as a baby learns to soothe itself. I wonder now if loving her made me worse.

Our naming of the fingers is, perhaps aptly, wrapped up in the act of destroying them. King Æthelbert of Kent first listed names for the five fingers in the sixth century when devising a system to compensate his subjects for accidental amputations.

"If a thumb be struck off, twenty shillings," he declared. "If a thumb nail be off, let bot [sic] be made with three shillings. If the shooting finger be struck off, let bot be made with eight shillings." And on he went from there: the middle finger worth four shillings; the "gold finger," six; and the "little finger," eleven.

Later, King Canute of Denmark developed a similar system of value for our fingers, but with different names. He, too, declared the most valuable the pollex, or thumb, and the least valuable was the middle finger, which he called the *impudicus*. *Impudicus*, meaning unchaste or immodest. The little finger, under Canute's rule, was called the *auricularis*, referring to the ear, because, at least in some theories, that finger is the perfect size and shape to remove the ear's sticky wax. The small finger later came to be called the pinky, it is said, because the pinky was once the name for a little boat in Scotland—the logic being, I guess, that pinkies are like little boats. Though what would this make the rest of the fingers? Tankers? Arks? Or, perhaps more fitting: pirate ships.

When I think about it, the desire began earlier than the biting. It started in second grade. One of my classmates wanted to teach me a magic trick. "Squeeze a blob here," he instructed, holding up the Elmer's and demonstrating on his palm. "Now let it dry."

We waited, waving our hands around in the air like acrobats. "Now," he said. "You peel."

And with this, he pulled layers of dried glue off his palm in large snowflakes. I followed his lead, exalting in the illusion of skin removed from skin, in the tickly feeling it created. The rest of the day, I tried the trick again and again in the back of the classroom, making and removing skin while everyone else glued Pluto and Mercury to cardboard rendered starlit. Desire preceded the act. There was satisfaction in so succinctly removing a part of what seemed like myself without pain.

The problem now is that it hurts.

This is not about self-pity—at least I hope not. It's about why any of us destroys our tools. It's about hands and what they are worth to us now. It's about the disconnect between our hands and survival, between our fingers and how we eat, between our thumbs and our reign over everything else living among us.

Isaac Newton once said that, absent other proof, thumbs alone should be enough to convince us of the existence of God. Perhaps fittingly, the thumb is also an appendage whose evolution no one can quite explain.

One theory is that the thumb developed after the threat of giant rodents forced our primate ancestors into the trees. Opposable thumbs helped those evacuees clutch branches, swing among the canopy, grab bundles of leaves for lunch. And when we eventually grew tired of the arboreal life and landed back on the ground, our new thumbs allowed us to grip tools, wield weapons, protect and feed our families and tribes. Our lives became less peripatetic, our tools more complicated, and our hands evolved with us: squat and square in early humanoids, longer and leaner in our parents and grandparents. And for most of these years, our hands were our most important tools; they were the weapons with which we battled the wild, the bowls and plates in which we cradled our meals. The phrase *living hand-to-mouth* originally characterized one major distinction between the rest of the animal kingdom and us. We don't bring our mouths to our food like dogs; we ferry the food to our mouths with our hands. Only recently has it been equated with poverty. You live hand-to-mouth when you are barely surviving. Something about our relationship with our hands has shifted. And here—finally—I am no longer talking just about myself.

One day, in the late 1960s, my mom decided to sign up for the air force reserves. She was young with long blonde hair, fingers stained yellow from nicotine, and nails bitten to the quick. A recent college graduate with experience as a chambermaid, she was directionless in that way many people are when they open the front door of a recruiting office. Vietnam was in full force then, so they should have embraced her, but something interfered. Fingernail biting is a sign of a personality disorder, they told her sternly, and turned her away.

Or so I remember the story. When I asked about it recently, though, my mom said she doesn't recall any of this. Perhaps trying to assuage, she added: "I do remember a therapist I went to once when I lived in Boston. I told him that my fiancé was really pressuring me to stop biting my nails, and I resented his implication that it was a sick habit. The therapist said, 'Well, it is a sick habit.' I never went back to him."

Another mom memory: I am very small and my mom asks me to help her weed between the bricks in our front walkway. "I need your tiny hands to get in the cracks," she says, and I look down, realizing that I, in fact, have tiny hands, and that, in this context, they are an asset; my hands are a tool.

Yet as adults we rarely contemplate our hands for longer than a few seconds. We look at them as we slip on rings, rub in lotion, grip a shovel, hold someone else's hand in ours. But it is rare just to stare at them. I can remember only one time I did so, at least in public, with other people around me. It was at a dinner party, and everyone at the table started talking about tricks of the body. A young guy there, fresh-faced, just out of college, halted the chatter simply by holding up his hands.

"I can't touch my thumbs to my pinkies," he said, modeling this disability, so odd that several of us protested:

"But wait!"

"No, of course you can."

"Just do this."

Suddenly we were all raising our palms up as if in prayer or praise, reproducing together the action he couldn't. We all stared at our hands then, thinking how easy it was for our little fingers to touch our thumbs, looking back at him with a tinge of pity, a hint of horror. In that moment I swear we all felt gratitude. For our hands.

I pause when writing this to trim the nail of my index finger with my teeth. The noise sounds like the knocking of tree branches against windowpanes. I pull and chew until I bleed. My thumb throbs, and yet I can't stop thinking about another bit of loose white skin protruding from a spot on my right pinkie that I tore at last night. I want to remove it, too. But I stop.

I am in seventh grade, and a girl named Brittany has everyone's attention.

"Did you hear she let a boy finger her during a movie in Mr. Adams's class," Amber whispers in the locker room as we change from jeans to sweats.

I already know about Brittany. She's tall and has long acorn hair that never looks fully brushed. Her chest is round and matches the pendulum swing of her hips when she passes me in the open-air hallways in our Florida middle school. She never talks to me, but I can't stop thinking about her.

Fingering was a word we learned in sixth grade. But this is the first time a story has attached itself to the act. Hearing it makes me feel uncomfortable—imagining a boy pushing his fingers into Brittany beneath the ambient light of a projector screen—and strangely nervous. Amber tells me the boy used two fingers, which also means something, I'm sure of it, though I didn't know what. Except that two is more than one but less than three.

What I didn't know then was that a year later I, too, would go the way of Brittany. Those were the '90s. I permed my blonde hair and started wearing shorts that stopped far short of my knees. I met a boy named Shawn, who told me I was sexy, which I decided was a compliment. I agreed to go to his friend Jason's house one day after school. And there, in a bathroom, Shawn touched me in the same way that boy touched Brittany in Mr. Adams's class. I didn't ask how many fingers. I didn't say a thing, in fact. Not even "this feels good," because it didn't. It felt gross, but it also felt grown up.

And several years later, I knew what it was like on the other side. Lesbians, many of us at least, make love predominantly with our hands. We eye one another's fingers, the strength of wrists, the muscularity of a thumb with an unspoken foresight, as if divining an experience in bed through the curve and bend of an index finger, a middle finger, the *impudicus*. Of all lovers, then, we are the ones who should be the most fastidious with our hands. I know this.

The woman I am dating now used to do palm readings. She tells me she has forgotten the details, so I try to refresh her memory with facts.

"The life line is the first formed in the womb," I say, holding her palm up for inspection. We are alone in my kitchen, miso soup reheating on the stove. "And then the destiny line. It's second."

She holds up my palms, inspects them, tracing these pathways, which, on my hands, ramble and digress but on hers run deep and uninterrupted from wrist toward thumb.

"In primates the heart and head line run as one," I say. "We are the only ones with lines that separate."

This fact seemed momentous to me when I'd read it a few days earlier. There is something special about us. Even without believing in God, I can see this. If it's not our bodies, then it's our ability to name them, to turn ordinary creases into metaphors and morph those metaphors into reasons to believe.

She holds both my hands now, and we look up, our fingers tracing each other's palms, a slip of space between us. Closing my eyes, I can feel the spots on my fingers where the flesh is broken and raw.

"You know, only the foot has more sweat glands than the hand," I say.

The soup has begun to boil, and its rumble soothes me.

"And at night the glands in the hand shut off. It's the only part in the body where they do that—"

One of her hands leaves mine and traces the side of my face.

"—where they go to sleep."

There is no space between us now, and my eyes are closed, but I keep thinking about hands. What I don't tell her is this: in the womb, babies develop nails in as early as twelve weeks, and, although this is thought to have an evolutionary benefit, it can also be dangerous. Every so often a baby is born with a battlefield of scratch lines across its face.

There are, of course, places for wolf biters on the Web. I visit these sites and flinch. All the while trying not to compare myself or think: I am better; I am not that weird.

"This is how I keep my Dermatophagia at bay before a party," a woman writes beside a picture of one hand, pink-painted nails peeking out from the Band-Aids wrapping each of her fingers.

"i wish i wish upon a star for SOMEONE TO GET RID OF THIS STUPID OBSESSIVE-COMPULSIVE DISORDER THAT MAKES MY HANDS UGLY," writes another. They hashtag *wolf biter, picker, self-harm,* and the technical term, *dermatophagia.* They talk about using their teeth, their fingers, and dad's Stanley knife.

But, every once in a while, you run across the rare unashamed wolf biter: "I bear these fingers with pride. I don't want to stop, I like the scars. I like the stares. It makes me unique," writes a woman whose photos are among the worst. Red fingernail polish half-chewed away. The flesh around the nails looks like snakeskin. I close the window and return to the page.

In 1908, doctors characterized nail biting and "finger picking" as a stigma of degeneration. In 1931, they called it an unresolved Oedipal complex. In 1977, a team of researchers released a paper on the "Relationship of Nailbiting to Sociopathy," in which they determined that sociopaths are more likely than the sane to bite their nails. Only, there was a hang-up: biters have long been thought to suffer from acute nervousness, and psychopaths are considered the calmest folks around.

In recent years, researchers have coined a new name for wolf biters, or at least for the action of biting: Body-Focused Repetitive Behavior.

"Some have theorized that there may be the same out-of-control grooming mechanism in the brain that underlies them all," psychologist Fred Penzel wrote in a 1995 article on skin picking and severe nail biting.

The most credible explanation to me, though, shows up only briefly in the world's first textbook on the subject: *Fingernail Biting: Theory, Research, and Treatment,* by Norman H. Hadley. Forced stillness, especially in the context of the classroom, Hadley writes in his summary of popular theories at the time, is one explanation for why we do what we do to our hands. "When we're forced to sit still we have to reassure ourselves—play with our hair, scratch or rub our skin, bite our nails—provoke sensations that keep us aware of our body."

In other words, we bite our fingers because, in stillness, we have no other way of proving the reality of our flesh. To stop biting, then, we would need to find another way to remind ourselves that we exist.

I read an article once about people suffering from Lesch-Nyhan syndrome, a debilitating affliction in which the patients, among other things, try to gnaw off their own hands. Doctors now believe there is a self-destruction instinct in all of us that is ordinarily topped by a stronger will to survive. Except in these people.

Perhaps, by this same logic, a wolf biter's survival instinct is on overdrive. We want so badly to feel alive that we remove pieces of our own flesh seeking proof. Then Newton would be wrong and Sir Charles Bell only half-right. The hand doesn't prove God; it proves us.

I took an art class recently, and the teacher told us: "You must be able to draw your own hand. This is where we all start."

Listening, I stared at my hands and saw the skin peeling from fingertips, scars along the flesh edges. My ring finger bore a notch along its tip, a chasm in the fingerprint swirl that recalled farmers market beets chopped too quickly, their magenta flesh fusing with my blood in a wound that had healed only a day at most when my teeth started ripping at it. On my index finger was a rosy fissure, the remnants of a hangnail torn from its base—again by my teeth—that had rooted along the seam of my fingernail. Each finger bore a similar scar, so that, at that moment, only my left pinky remained unscathed. I looked at all of this as if it weren't me, and then I began to draw, starting with my thumb and moving to my fingers, trying to copy perfectly the ridges and swells.

And for that moment, at least, I was still, making a replica of myself.

HESTER KAPLAN

The Private Life of Skin

I.

"THE SKIN'S THE best part," he said.

I knew that. The skin is an exquisite organ, elastic, intelligent, highly sensitive, strong-willed, and always honest.

"Juicy and sweet," he added, and winked at me like I was a virgin. "That's because there's lots of fat."

This culinary dermatologist in a grease-stained T-shirt was talking about the 130-pound pig rotating on a spit over a bed of coals the size of a twin mattress. Determined to make his point, the man hoisted himself out of a sagging lawn chair and punctured the pig's skin with a long-handled fork. There was a slight hiss, like the sound of a deflating balloon, and then juice began to weep from the two holes like high-cholesterol tears, and sizzle on the fire below. The smell was salty and enticing, bacon's pheromones at work on me. The man poked again, wheezing as he wielded his weapon.

He was a professional pig-roaster, hired by my sister on the occasion of my father's seventy-fifth birthday. The roaster's own skin was stretched to a tight shine and was the alarming color of strawberry Kool-Aid, a result, I guessed, of the heat and the fact that he'd been sitting in an upstate New York field, in late July, without hat or shade since early morning. He was enormous; even his fingers appeared to throb with hypertension. And then there was the bed of his pickup truck that was full of empty Schlitz cans, half of which he'd used to baste the pig, the other half to baste himself. He was the coming attraction for his own catastrophe.

Dinner—this glazed, dripping pig—was way past catastrophe. Eviscerated but not beheaded, it clung to the spit as though it had only intended a climb on the jungle gym, and having once glimpsed the peril of forks and knives, three kinds of mustard, and sandwich rolls waiting below, had decided not to come down. The animal's eyes were greasy black marbles that resisted my attempt to read anything like despair in them. Its skin had gone from a fuzzy, babyish pink (or so I imagined) to a deep rusty and fissured golden brown.

"Male or female?" I asked.

"Never looked," the roaster said.

There was nothing to see down there anyway except a large, loosely sutured wound. Gender in pigs was apparently without its own specific flavor.

Far behind me up the gentle hill, despite the fiercely glaring sun, the large party went on, determinedly festive under the beautiful swoop of a white tent. I didn't see my father anywhere. In spite of the alarming heat and humidity, I was in long pants and a shirt with long sleeves, hiding my troubled skin from view, hiding myself way out here by the doomed pig.

"Taste?" the roaster offered.

I nodded and looked again for my father. I knew he would want to try this as much as I did, but apparently he was hiding, too.

II.

A month earlier, my psoriasis, which I'd had for almost two years—the plaque variety that was more a shiny red than flakey white—had taken a turn for the worse, covering an alarmingly large percentage of my body. When I thought about what was happening to me at forty-two, I thought in terms of expanding armies and colonization, the brutal, in-the-dark takeover of previously bucolic fields and peace-loving peasants. Still, it had been a shock to hear my doctor say that I was now a candidate for day hospitalization, a shock to learn my condition even warranted that consideration. It was like going in to have a cavity filled and having the dentist suggest dentures. My dermatologist had a habit of smirking that I'm sure he thought looked like sympathy. The view from his office was of Route 95 and the city's newest tourist attraction, a decommissioned Russian sub. I thought about stopping there for a visit if I ever got out of this place.

Until then, for the past eighteen months, I'd been happy enough to get up at 6:00 a.m. three times a week, even in the brutal Rhode Island cold, and drive to the hospital for phototherapy treatments. It was a private errand very few people knew I made, my children still asleep when I left the house, my husband just stirring. I loved the empty streets at that hour, the chance to see the sweet and troubled city without traffic, and there was always something beautiful to come across—the perfect cube of a cardboard box blown into the middle of an intersection, enticing like a gift. In the same way that some people write their novels before anyone else is up, or go running before the newspaper's even been tossed onto the porch, this was my early morning ritual, essential and virtuous. Granted, it was a strange ritual in which I got completely naked (I sometimes thought I should drive over in my bathrobe and save myself the trouble), covered my nipples with sunblock, my face with towels, and my eyes with white goggles held on by an elastic strap, and stepped into what looked like a ceiling-less phone booth lined with long, vertical light bulbs. Sometimes a nurse on the other side of the booth, before she set the machine to zap me with UVB rays, announced, *I'm going to look now,* which was my warning that she was about to open the door and assess my skin, measure my progress—or not—toward clearing. I'm not sure if it's more uncomfortable to be scrutinized naked when you can't see the other person's face because your own is hidden behind towels and ridiculous goggles, or when you can. Here the shame was multiplied in a hundred ways. Her touch, her gentle pressure to turn me left or right, always startled me.

Sometimes during my two-minute treatments, I listened to the conversations going on between the nurses and other patients who came and went as I did. They discussed the weather most of the time, as did I, poignantly benign talk between people who showed everything here but knew little about each other. Standing in the booth often lead me to imagine that the lovely warmth I felt was real sun on my skin and not some highly calibrated pretender okayed by my insurance company. Phototherapy still struck me as the most primitive of treatments, just light after all, what we've always had, what we've always sought. This was the tanning salon without vanity. A computerized male voice informed me when my session was over and asked that I "please leave the booth." Dressed and back in the hospital hallways, elevators, and even in the parking lot, I could always detect my comrades-in-skin because they were the ones who, like me, smelled of sunblock, summer, and the beach, a hint of coconut oil, even in the middle of winter.

But the idea of day hospitalization (enter after breakfast, leave before dinner) to get me over this "hump" meant things were much worse than I'd wanted to admit, and that my morning dose of rays was no longer working. That it wasn't enough. In some places my skin was so raw it was as if I'd been seared by a blowtorch—my ass, for instance, which made sitting down its own peculiar punishment. My condition, until then the equivalent of a despotic hobby, was about to become a full-time job. The hospital would be my office. I would bring my laptop, my lunch in a brown paper bag. Would I take calls while sitting in a tub filled with foul-smelling potions? Could I hold the phone while I was smeared with tar ointments? Would I have colleagues, all of us meeting in our gowns and disposable slippers to gossip and trade advice and bologna sandwiches? Oddly, one of the books I was ghostwriting at the time was about skin cancer, and on the morning of my appointment with the smirking dermatologist, I had taken too long with a single sentence, not because it was difficult or complicated or even particularly smart, but because it needled me in a way that others hadn't.

"Through a language uniquely its own," it read, "the skin communicates information about each of us—how we live, who we are, how we feel."

I'd always known this was true about my father, an enigmatic, mostly evasive, painfully private, life-long sufferer of eczema. *Keep it all in,* I'd thought smugly, *and this is what happens, this is how you're punished.* His skin was how I knew him, after all, how I'd always known what he was feeling and recalling when he was silent and gave nothing away. But now I wondered, how could any of this possibly apply to me? I was not my skin, that ravaged and distressed piece of clothing, *I* was not ravaged and distressed. I told the doctor I'd think about it.

III.

My father's song went like this: itch, scratch, rake, gouge, itch, scratch, bleed, sigh. He could play it with one hand, fingers trilling in a restrained crescendo over a patch of inflamed skin, or with two hands in a rousing chorus of atopic agony. Like the singer who belts out "America the Beautiful" at the start of the talent show, my father delivered his anthem with unabashed fervor. He was a citizen of the country of eczema with its silver-crimson fields, its fissured rivers, its purple valleys of irritation. And like the true patriot, he found it impossible to imagine living anywhere else.

Other people remember waking in the night to the noise of their parents fighting or having sex, or roaring downstairs with the neighbors, one of whom, it may turn out years later, was a lover. But I was nudged awake by what seemed like waves seeping onto the sand—here, in a house in the city, the only nearby water the fetid Charles River—but which was in fact the sound of my father scratching himself in his sleep, the susurrus of fingernails moving across skin. The dog and the humidifier wheezed out in the hall, the refrigerator rattled, and my father somnolently answered the urgency of an itch. In the morning, I sometimes heard my mother mumble when she made the bed; the sheet on my father's side was a galaxy of bloody stars and comets, and hers was the expression of a slightly exasperated, but loyal astronomer.

My father's scratching was not just a night event; he was no nocturnal creature like the opossum that sometimes crossed the moonlit yard. There were times during the day when he could not still his hands, could not keep them from galloping across his skin while he wrote in his office on the first floor of the house, or stirred soup, or made sarcastic comments to me as we sat through "Love Story" at the Fresh Pond Cinema while everyone around us wept. He could not stop them after a shower when he hung around in his green terrycloth bathrobe, or when he drove his three daughters to school, or argued with the man who'd come late to fix the washing machine. He took walks with Olive, the ancient beagle, by the malodorous river and the two of them scratched there, I'm sure. Only in Cape Cod Bay, with its curative waters, where he liked to float on his back, his arms flopping around like spastic wings, were his hands really ever off duty, was his skin ever at rest. The gulls above watched him curiously, this pale, flailing fish. Occasionally I'd catch someone—one of my friends or one of his, or a stranger in line at the A & P—watching my father go at his inflamed and cracked skin, and I would allow myself for a second to see how tormented he must have looked to them.

Sometimes the noise and constant movement of his scratching drove us crazy. "Stop itching!" we'd scream, using the single verb that managed to encompass action *and* sensation. He rebuffed us only for what he called our "female shrieking" and not our impatience, which he ignored—or maybe even found amusing. His older brother, imposing and impressively bald, once scolded us both for being dopey with the English language and for our lack of compassion. "Eczema is a disease," he whispered, "and your father is a victim. Do you tell a cripple to stop limping?" he asked, himself a cane-carrying casualty of childhood osteomyelitis. Behind my uncle that day, my father shrugged and raked the skin between his fingers. No doubt being called a victim, and having his condition labeled a disease, made him uneasy. He didn't think of himself that way at all; he was not like his brother, despite the fact that his affliction was just as obvious. And he never spoke of it in any deep or personal way. Eczema was not something that happened to him, it was him, and he'd had it all his life. He once told me he thought he'd been born itching and scratching, maybe even *in utero*.

The pig roaster leaned toward me. "I'll slice you up a taste," he said.

For all his heavy breathing and cinnamon face, he was adept with his knife and sliced off a narrow band of skin, five inches long, which he then cut in half and laid on the white paper covering the table. The skin retained some stubborn life and curled up as though trying to escape. It sweat its oily ghost onto the mat. When I put the piece in my mouth, it exploded, delicious, seductive, sweet, salty, a balm of fat to soothe the conscience, a little crunch of resistance. The roaster popped open another can of beer and poured it over the carcass that sighed in cool relief like any overheated inebriate. The man's eyebrows were raised expectantly while I chewed. I wanted my father to try this—the revenge on skin would not be lost on him—but I still didn't see him under the tent or anywhere among the drooping guests.

"Didn't I tell you it's the best part?" he asked.

I nodded, but sometimes the skin's worst part, too, as it happened to be at that particular moment. For several days leading up to the party, I'd felt the ominous, unmistakable tingling, the creeping, frustrating, fear-inducing ache of an approaching flare-up of my psoriasis. It was what the doctor had warned me about with a thoughtful tilt of his head. It wasn't there yet, but it was coming for sure, and I already knew that the pig and I were cooked, each in our own way.

IV.

When I was about twelve, four years older than my father had been when both his parents died, his scratching began to embarrass me. It made him seem strange, possessed, maybe even a little freakish, like the dwarf at the end of the block, or the boy in my class who had a tic that made his lip lift and quiver. I started to wonder what would happen if my father denied his urge to scratch. What if he were tied up in a chair, his hands bound in tape so that he couldn't touch himself at all? It would be a very cool experiment, cruelty aside. Would his head explode? Would he have a fit where his eyes rolled back, he panted, drooled, babbled until he passed out? Once the crucial, excruciating moments had passed, when his body realized it would not be relieved of its itch and it would survive, would the impulse have evaporated? Would he be cured?

Short of that, one afternoon I braved interrupting him in his office—something not done unless the matter was urgent, a fire in the kitchen, or the dog escaped down the street—to tell him a joke. It wasn't particularly for laughs because behind it was a lesson, a suggestion. This was the way my father and I talked to each other, mostly through jokes, obliquely, teasingly, a swipe at the surface of sentiment.

It was the rare occasion when I heard my father talk about what he was feeling. He would offer a literary passage in lieu of confession, he would much rather quote you a line of poetry about sadness than ever speak of his own. And while I had not yet discovered what might speak for me in this way, I had already begun not to speak about myself. A joke would do.

When I went in that day, he didn't put his book down but shut it, using his finger to hold his place. Okay, I said, so. He scratched his leg. So. This family brings the father to a doctor because he can't stop eating paper. It's a problem. They've tried everything—shrinks, drugs, shock therapy, punishment—but nothing works; the man still eats paper. The doctor takes the father into his office and at the end of the visit brings him out and announces that he's cured. "How did you do it?" the family asks, amazed and grateful. The doctor shrugs; he's a little bit of a genius and a schlub. "I just told him to stop eating paper."

"Hah, hah, very funny," my father said.

I looked at his finger still holding his place. My face burned.

Was I hinting that he needed to be cured of something and that all he needed to do was "just stop scratching," that it was a matter of will? What exactly had I wanted to ask him that day? *Who are you, what do you feel? What was it like when your parents died? Why is your skin so terrible? Why don't you cry at movies?*

Thirty years later, with my own skin affliction, I still like the joke for the beauty of the idiot miracle cure and its loopy suggestion that the psyche, after all, is not really such a hard nut to crack; just say no to it. If only. I have been told by the many doctors I've consulted and pleaded to for help, that my psoriasis, like my father's eczema, is exacerbated by anxiety, stress, self-consciousness, tension, worry, and I always think, well, what besides death isn't? The heart beats faster when we're scared, the chest clenches as we dial 911, the stomach flips with remorse, the head pounds with indecision, the mouth waters for a kiss; we are our bodies. My father knew this, and his skin was a billboard for his life's vicissitudes, the most public expression of the most private man. What else did he need to say?

If he could not, or would not, talk about his sorrows, his tormented skin and his scratching at it gave him away. But the bind was that while scratching brought glorious relief in the moment, it only made his skin worse later. As a child, he was constantly told to stop scratching, which only added rage to the itching cauldron. The act itself was deeply and defiantly powerful in its ability to soothe, to comfort, and to assert himself, and at the same time, deeply weak for its inability to resist what was harmful. But he scratched because it reminded him exactly who he was.

I've read that people can stand pain; it is the itch they cannot endure.

v.

In time-elapsed photography, crocuses nose their way through the dirt and bloom in seconds, the tadpole becomes a frog, a ranch house a mansion, my boys' upper lips darken with hair.

Time-elapsed photography has the best elements of any story—
expectation, suspense, economy—and I knew that if I could tell
the (very short) story of my own skin disease like this, I might
someday hit upon a satisfying ending. Its start occurred over the
course of about ten months when I was forty, when I had three
bouts of strep throat, each worse than the next.

Any number of factors can initiate psoriasis or cause flares,
and an infection with the streptococcal bacterium may be one
as its toxins activate the skin's immune cells. The pathogenesis
of my problem was interesting and useful only to a point; what
was more compelling was what I could see happening. Three
or four spots on my legs, discrete, red, and rain-drop sized, became eight and then ten. More bloomed on my back, chest,
and behind my ears. Soon my front was covered with a herd
of the spots, migrating wildebeest. Creams only made me burn
and rush to the cold shower. The spots became ravenous and
disoriented, driven by a hunger for clear skin where the feeding
was sweetest—valley of inner thigh, cave of underarm, river of
C-section scar. I learned beautiful new terms like axillary and
popliteal. "The trick," one doctor advised peremptorily, "is to
not think about it too much." *Stop eating paper.* I wanted to
tell him to go fuck himself. The trick, I thought, is not to think
about *anything* too much.

My father once told me that the worst thing he ever heard
from a doctor was when he was about twelve. "The man looked
at my raw and abraded arms and said gleefully, 'I see you've
been having a good time!' I wanted to kill him."

VI.

Just before his first book was published, the one he'd been
working on for what seemed like most of my childhood, my
father suffered a tornado-sized flare-up of his eczema. We'd
been living in England for the summer in a rented place called
the Oast House, and he spent those glorious, boxwood-scented

days correcting his galleys that arrived in long ribbons of paper and hung off the edge of the table like exhausted tongues. He sat in the oast room itself, an empty octagonal space that was flooded with a high, thin light. Across the road was a field of appealing cows and charging, snorting bulls with fascinatingly large, swinging balls, and the scent in my father's makeshift office was part manure, part newly mown hay, and part anxiety. On the floor, pink eraser rubbings, like tiny curls of skin, surrounded his chair. I'd seen this happen at school during tests, the surest sign that a nervous kid was about to lose it. At the end of the day, my father emerged into those violet English evenings I've still never seen anywhere else, glassy-eyed and scratching like crazy. He drank a gin and tonic. His arms were gears, windmills moving around him. His skin was purple, ominous, the color of a bruise that must have hurt a lot. We took one look at him and ran outside, lured by the scent of the fields, the bulls. A few days after we returned to Boston, my father went into the hospital.

At the time, he was under the care of a world-famous dermatologist, an Irishman in a field dominated by Jews. (In college at fifteen, he'd been seen by a doctor with the distressing title of Professor of Syphilology.) In a house where no one—particularly professionals who took themselves too seriously—ever escaped harsh judgment and some measure of ridicule, this man was untouchable. It was this doctor who declared that my father's flare-up was caused by acute anxiety over finishing his book. My father knew exactly what could happen to a book—nothing, and sometimes worse than nothing. The doctor admitted my father to the hospital and told him to pretend he was on vacation there, and apparently that's what he did, inviting friends in to sit with him in the evenings, while he was swaddled in plastic wrap and basked in the fuzzy, swollen warmth of steroids.

This wasn't the first time he'd been hospitalized for his eczema. Stories of his earlier stays (some before I was born) always had for me a kind of suave bachelor feel to them, as though he were talking about jazz clubs in New York, or beautiful women he'd dated, or smoked salmon omelets he'd eaten at midnight. I pictured nurses like stewardesses, their wrists cocked suggestively, offering mixed drinks, not bedpans and apple juice. In his private room he would read, listen to the Red Sox on the radio, be away from his family, and have his meals brought to him. In the white light of Massachusetts General Hospital, now as before, his skin would heal.

Around the time I was felled by strep infections and saw the onset of psoriasis, I'd had two books of my own come out. Reviews kept me in a state of nervous expectation. I thought about how every year there was more I wanted to get done and how it might not be possible. I imagined I knew, in some small part, what my father must have felt like that summer.

"I loved being in the hospital," he said. "It was a terrific time, crazy as that may sound. I thought about absolutely nothing."

Under the tent, the pile of gifts grew as more people arrived at the party. There was a bag containing Preparation H, a pair of drugstore reading glasses, a four pack of Rolaids, Aleve, and Metamucil. There was a red hot-water bottle in the shape of a devil's horned head, a jar of Tucks Medicated Pads, and another tube of hemorrhoid cream. There were the requisite CDs (mostly "Best Of" collections) but the theme here was obvious: decrepitude and iffy bowel function. It was funny but also bittersweet, true to the essence of any birthday party. Behind the boozy, generous faces, between the margaritas and the chicken wings, we all think about our own slide toward the end. Not in a panicked or morose way, but mortality flies across the horizon. For an instant, there it is, your life, and then gone so quickly you might think it was just a bird passing around the pines.

I knew it was time to bring my father out here, to let him take a bite of the delectable skin. I wanted to ask him now too, if I could have found him at his own birthday party, if what he'd liked about being in the hospital years ago—this thinking about nothing—was what we were really after. Was it thinking too much that got us into dermatologic trouble? Or was it the not talking? But the truth was, I wouldn't have asked him any of this, even if he'd appeared just then. It wasn't what we did.

VII.

In high school, I had a boyfriend who liked to stroke my cheek. He said I had beautiful skin, which made me turn red instantly. I hated how easily I blushed; it seemed cruel that despite my efforts to be unreadable, what I was feeling was always clear on my face. I wasn't cute in the way my friends were, and I didn't possess their kind of perkiness or comfort in my body, but pimples were never my agony. But this boy's touch with his long, thin fingers confused me. At sixteen, it felt too old, too much about remembering something that would be distorted and wrinkled by years, when we were right there and young at that moment. At the very least, we should have been grinding against each other in the backseat, not conjuring nostalgia that didn't even exist yet. I thought about this not long ago when I read in my alumni magazine that he'd died of complications from AIDS. I'd heard years before that he'd come out during his freshman year in college, and I'd hoped then that the particular tenderness he'd shown me, that tender, melancholy confusion, was met with appreciation. His gesture, back of hand against my face, now seems like a kind of sensuality that was beyond sexuality, beyond anything he was able to express. Certainly beyond anything I was able to understand. It's a move I now sometimes find myself making with my sons, touching their smooth and beautiful faces, even as I anticipate the appearance of acne and beards, and they push me away because mine is not the hand they want anymore. I recognize it now as gesture that is ripe in its prescience of inevitable loss. The skin is the first thing we mourn when we age because it is the first thing we see when we look at ourselves.

VIII.

For a while, I eavesdropped on an Internet chat room for people with psoriasis. I knew almost immediately (given that I only "listened" and never "spoke") that this wasn't the place for me. There is something too binding about this kind of primary self-definition—why not curly-haired people, or historic novel enthusiasts, or salt-and-pepper-shaker collectors instead who also happen to have a skin condition?—but I wanted to know what other people were feeling. Was I normal to be panicked on some days, in despair on others?

I talked to my husband, but I didn't talk to my friends or my sisters or my parents about how truly distressed I was, or how I teetered on the edge of depression. I was embarrassed, humiliated by how I looked, full of self-loathing, and I was afraid that if I exposed myself, physically or emotionally, I would be pitied, scrutinized, misunderstood. I'd look like a mess. I'm sure it appeared that I was handling things admirably, without fuss, and why not? My face and hands and forearms had been spared, so what was ugly was covered. But I showered with the lights off and cried alone in parking lots. I felt I'd been abandoned by my own self and invaded by another. My father occasionally asked how things were with my skin, and I always answered in a clinical way. Sometimes I told him funny stories about phototherapy and the kindergarten-sized locker I had there. I wasn't dishonest, but I wasn't particularly honest either with the one person who might have given me some understanding about the peculiar confusion of appearance and self, of disease and psyche.

There was a certain irony that an online support group for people with a skin condition—which, among many things, is a problem of self-presentation—should involve people who never "saw" each other. Many members seemed to spend a lot of time in the chat room. This enthusiasm itself seemed to carry a poor prognosis, the equivalent of wanting to be good but still hanging out with the bad kids at school. *Haven't been out romantically in fifteen years; how could I ever get naked in front of a man? I don't go to the beach because people stare at me; I read travel books instead. I attend an online university; I should graduate in 2009.*

I followed the drama of B., a man who happened to mention, in the course of a longer story about a new medication he was trying, that his fiancée had called him Lizardface in front of the wedding planner. The regulars instantly weighed in, and the exclamation point got a serious workout.

Dump the bitch! You deserve better, no one understands what it's like but fellow sufferers! Learn to love yourself for who you are! Who the hell needs her!

(My husband called me Spot and I didn't mind at all. He lay beside me every night and I never felt ugly then.) B. seemed genuinely surprised by the vitriol from these people who spent so much time bemoaning the fact that they were alone, the same people who on other occasions had told him how happy they were he was getting married, as though they were sending one of their players off to the big league. I began to imagine the chat room as a white space with buzzing lights, a plastic pitcher of warm water, and people comparing their plaques, their ointments and miracle remedies cooked up in the back rooms of quacks and healers. (If you look online for these, you'll find plenty.) I couldn't get out of there fast enough.

My father would have understood; there's a lot to be said for suffering alone.

IX.

If you have psoriasis, you're something of an idiot savant when it comes to skin. You know a lot, but you know almost nothing useful. A million things can trigger a flare-up, but because there's no predicting, there's also no way to prevent it; this is one of the golden nuggets of truly useless, essential knowledge I've acquired. Some contributing and exacerbating factors beyond stress are agreed upon—meaning they are what the doctors know and will tell you about—but also what you might find in any waiting room brochure: weather, change of seasons, heat, cold, moisture, dryness, ibuprofen, allergies, alcohol, sugar, a cold, a scrape, obesity, the antimalarial medication chloroquine, streptococcal infection, illness, menstruation, trauma. But there are many more unknown factors, and everyone with psoriasis has his or her personal list, though half is purely fantastic. My own list on that day of my father's birthday party: teenage children, a short story I was struggling with, a husband hit hard by the death of a beloved brother-in-law, that bird passing behind the pines.

In this disease, the arts of medicine and superstition overlap, and pathogenesis is sometimes just another way of telling a story, no more or less believable. People with psoriasis look at the world with a cautious eye because they know that every event, every emotion, every everything—word, bird, breeze, spice, memory, curl of pig skin—might soothe, exacerbate, erupt, tease, or destroy the skin. If we once believed we were in control, at least to some small degree, of how we responded to life, disease is a constant reminder that we were wrong. In psoriasis, the skin cells often multiply ten times faster than normal. So many of them rush to the surface that they get jammed up like commuters trying to get out of a train. What's your hurry? I want to scream. We have lots of time, so much time!

The pig roaster, head heavy on his chest, was either drunk or dead, while his charge turned on the spit. I felt my skin begin to effervesce in earnest then, starting at the small of my back, running down my back and my legs. It tingled across my stomach. Soon I would be covered in tiny bumps as though I'd been splashed with sand. And not long after that, the bumps would explode as I looked on helplessly like a child locked inside during the fireworks.

There was no guarantee that day hospitalization would actually help. I could go on hydroxyurea, my doctor had said, explaining that it was a chemotherapy drug and I'd need to have my blood taken every week to monitor my white cell count. Pregnancy, he added, looking at the floor, would not be a good idea.

x.

Who doesn't come to their own party? People stood at squinty angles with their hands visoring their eyes like passengers on a cruise ship. My husband's posture was slightly defeated by the pressing humidity, my mother's laugh a little tight as she held onto my brother-in-law's arm, her eyes hidden behind Sophia Loren sunglasses. My skin was suddenly on fire and my joints felt swollen and stiff. The heat was unbearable. I wanted to rip my clothes off—but wouldn't everyone see then? I thought of all the things I'd like to dip myself in at that moment: ice water, cold, heavy cream, lime Jell-O. It was time to find my father.

It was blissfully cool inside my sister's house and I felt my skin instantly relax. Beyond the center stairs, I saw my father's feet slipped out of his sneakers and heard the steady whisk of his fingers across his skin, a song as familiar as "Happy Birthday" which we'd all sing soon enough. He'd known enough to be in the air conditioning all this time while I'd been standing in the heat, by the fire. He knew what he needed to be comfortable. The television was on, I assumed, only because there wasn't a book or a magazine my father could have picked up to read instead. He looked up at me sheepishly.

"Tiger's playing," he said, nodding towards the screen. He had no interest in golf, but appreciated the narcotizing effect of all that blemishless green.

I sat next to him. "I stink like pig."

"That's not a nice thing to say about yourself." With his left hand, he raked his forearm and then he went for the back of his neck. His skin had the flaky and silvery shine of a piece of mica. "Actually, more like bacon."

"Stop itching. What are you doing in here?"

"It's too hot out. Admit it, isn't it nicer in here?"

"This is your party and you're hiding."

"I'm not hiding. I'm just being by myself. There's a difference." He shrugged. "And I was thinking that I'm not old enough to have a seventy-fifth birthday. So I've decided it's a mistake. The real birthday boy just hasn't shown up yet. Let him blow out the candles. I'll watch golf."

"I'm considering going into a day hospital program," I said after a while. "Or start on this nasty, toxic drug that produces two-headed babies."

We hadn't been talking about my skin, but he knew what I was saying; that it was worse. He didn't look at me. "What do they say?" he asked.

He knew, of course, that "they" never said anything very useful, that there was no cure for what either of us had. Only management or lack of management. Luck or lack of it. Acceptance or not.

"What they say is the worst thing you can do is think about it too much. But I doubt that's really the worst thing. Taking a piece of sand paper to your skin would probably be worse. Or maybe a file."

"Or a cheese grater," he added. "I used a hairbrush once—that was pretty wonderful—and a comb another time when I couldn't find the brush." He held up his hands to show me his nails, kept short on purpose. "Disarmament."

"I'm hoping this will just go away on its own." Why was our talk still surface, when what I most needed was to say and hear what was underneath; how was I going to live with this? I wondered if we'd ever had a real talk.

A player hit the golf ball into the water with a brilliant splash, and after a few minutes, my father said, "I'm sorry you inherited this from me, this troubled, terrible skin. I'm really very sorry."

"But they say that's not how it works." His face had fallen and it pained me to think he felt responsible. "Eczema, psoriasis, they're not related at all."

"But you know it's true, and so do I. How could it not be? And 'they,' by the way, know almost nothing. It just makes sense. You're like me. You keep what you're thinking and feeling to yourself, and that's how you want it, that's who you are. You know it's all there though, bubbling around and making up stories."

"I'm not sure that's okay. I'd like to be different."

"But you're not. And that *is* okay," he insisted. "It's just that sometimes your skin rats you out. Nothing you can do about that."

What he'd said was true. We shared the dual inflictions of painful skin and painful privacy as much as we shared blood. I couldn't ask him about my prognosis, or about what I would write next, or about what it was like to have your children grow up. But maybe on the occasion of his seventy-fifth birthday, this was what we could do, share some sense of the mystery of our strange selves. This was the conversation of concealed people, though he never hid his skin from the world. Sometimes those who are most hidden, those who can't talk, are revealed in this way. What we want to keep inside is bared on skin that's only ever been seeking solace, company, the sun.

XI.

The pig roaster wrenched a noisy leg off the animal. He was sweating great drops, his complexion growing darker. The meat itself, when it was finally hacked up and spread over the white paper, was a beautiful rose color in the near dusk, but surprisingly tasteless. I watched my father take another curl of skin, pop it in his mouth, and wipe his greasy fingers on his pants. He was grinning while he chewed. The birthday boy.

"Don't you think the skin's the best part?" I asked him.

He nodded and wiped his mouth with the back of his hand. "Grab me another piece, will you?"

Later, as the temperature and the light began to fall and the sky turned purple over the house, there was a cake dotted with strawberries and festooned with swags of red, white, and blue icing. My father swayed behind the candles to the singing of "Happy Birthday." We all wished it could be us serenaded at that moment and we were just as happy that it wasn't. My skin felt like champagne, worrisome probably, but at the moment, it was a bubbling up, an exuberant burst of mortality that sometimes arrives in troubled, emphatic, and humbling form.

WENDY CALL

Beautiful Flesh

A BIRD'S PANCREAS looks very much like ours does, slim along the intestine. Nearly every creature with a backbone has a pancreas, lungfish and lamprey eels and ray-finned fishes being notable exceptions. In mammals, the organ is always small, shown in anatomical drawings peeping from behind the stomach or duodenum of the lion, the rabbit, the chimpanzee: red in dogs, brown or green in frogs, pinkish tan in humans. The human pancreas looks like a fleshy leaf, like an ornament on a gilded painting offering glory to the gods. It reclines between duodenum and spleen, linked to both by elegant, fragile tubes, silent and unassuming until something goes awry.

This organ is not really an organ at all, but different kinds of cells living together. Doctors have known about the pancreas since the time of Socrates but for centuries had no idea what it did. Galen, physician to the Roman emperor, considered it a cushion for the blood vessels around it, with no other purpose. (Honestly, do blood vessels ever need a cushion?) Galen probably got that idea from Herophilus, the man who first described the pancreas. Herophilus made his discovery through a medical practice he had pioneered: public dissections of human bodies—perhaps even some not yet quite dead. Forbidden from continuing that tradition, Galen made do with pigs. Thanks to him, and to Roman law, the mystery of the pancreas extended into the twentieth century.

Two centuries ago, we called the organ "sweetbread." Aristotle used the word *pancreas,* which means "all flesh" in Greek, but he might have been referring to all the glands of the body.

(Translation can be such a pesky thing.) Galen called the organ he thought a cushion *kalikreas,* or "beautiful flesh."

After the medical profession discarded that cushion idea, doctors considered the pancreas a waste collector: pancreas was to spleen as gall bladder is to liver. However, a Dutch medical student, who would later discover ovulation, proved the pancreas produced an important fluid. Two centuries before anesthesia was discovered, Regnier de Graaf sliced open dogs' bellies and slid goose quills into their pancreatic ducts, so that the organ's juices could not slip through the sphincter of Oddi and spill into the duodenum, as they should. Those hollow quills siphoned canine pancreatic elixir into jars tied under the dogs' bellies.

Fortunately, de Graaf began his medical discoveries very young; he was killed by the bubonic plague when he was thirty-two. The year he succumbed to the plague, a Swiss doctor removed the pancreases of several dogs. Most of his research specimens ran away—they had been stolen and ran back to their owners—but those that stayed grew thirstier and thirstier, urinated much, and died within a few months. So much for the waste collector idea.

When I learned my mother had pancreatic cancer, I was insufficiently fazed. I try now to imagine what I was thinking, before I knew what I should have been thinking. Perhaps I thought, "Oh, cancer like breast" or "like lung." (I should have been thinking "like bubonic plague.")

In 1825 a New England doctor wrote down the recipe for an anti-tumor balm. *Simmer one part rosin beeswax and four parts mutton suet. Add one ounce lead to a pound of salve. Simmer until lead dissolves.* In the nineteenth century, treatments for cancer included hemlock, foxglove, belladonna, and blue rocket. All, like lead, are deadly poisons.

Other cancer remedies included directly applying to the patient's body gastric juices, ox blood, slices of raw animal flesh, iodine, and copper acetate. Some doctors tried to starve out cancers by restricting their patients to distilled water; others instructed their patients to swallow dozens upon dozens of skinned, gutted, still-palpitating green lizards. In 1835 a doctor dismissed those treatments as quackery, instead recommending creosote and arsenic.

My mother's oncologist wanted her to take Gemzar and Tarceva. We knew which of her fellow patients were on the first medication because of the flaming red rash that peeled thick layers of skin from their arms and faces; we recognized those on the second drug from the acne that bloomed on their middle-aged cheeks. They all died anyway.

Dr. Oddi, a late-nineteenth-century Italian physiologist—who dissected cats, chickens, dogs, humans, oxen, pigeons, pigs, and sheep, so he could draw their organs—compared the texture of the pancreas to raw silk. He became a depressive cocaine addict and was eventually tried in a court of law for killing one of his patients through overmedication. His name lives on as a pancreatic duct.

Things that your pancreas does not like: Vienna sausages, Round-Up weed killer, Coors six-packs, ten-dollar boxes of Inglenook, gasoline, aspirin, beta-blockers, cigarettes, hundred-dollar bottles of pinot noir, horse kicks to the belly, petroleum spirits, scorpion stings, dry-cleaning chemicals, salami, steering wheels in high-speed vehicle collisions, agricultural pesticides, anti-psychotic medications, bright red dye, and strong coffee.

Don't fuck with the pancreas: That's the advice my mother's general practitioner received in medical school. That same doctor gave Mom a clean bill of health three months before an oncologist gave her a terminal diagnosis. Five years before that "you're-fine-oh-no-you're-dying" chain of events, my mother had abdominal surgery at a military hospital. A troop of new Army doctors was paraded into the operating room to see my mother's belly-parts splayed; they were invited to take probes and peek inside. My mother believed that someone, quite innocently, fucked with her pancreas. She felt no animosity toward that person, saying, "Well, they have to learn how to become doctors somehow."

Before my mother's stage-four diagnosis, the only thing I knew about the pancreas was that if it didn't work properly you ended up with diabetes mellitus. *Mellitus* means "sweet like honey," which describes the taste of a diabetic's urine. Before insulin was available, diabetics were treated by limiting the carbohydrates they ate and feeding them fresh veal pancreas. Thirty-five hundred years ago, Egyptians wrote the symptoms of diabetes on a papyrus scroll, part of a medical treatise more than sixty feet long. In the 1870s, Georg Ebers, romance novelist and Egyptologist, found that papyrus scroll in Thebes and published a German translation. A surgeon who lived near the Euphrates River, Aretaeus, named the disease in the first century, describing it as a "liquefaction of the flesh and bones into urine."

To the surprise of no one but my immediate family, not long after my mother was diagnosed with pancreatic cancer, she became diabetic.

For twenty-two years, until my mother married my father, she lived in a tiny house on Rust Avenue in small-town Michigan, with her mother, father, and older sister. For twenty-two years, five days a week, my grandfather came home from his job at the Big Rapids Gas Company, his work clothes laden with chemicals. After my mother married my father, they moved to a Southern California agricultural town. Several times each week throughout the long growing season, a crop duster buzzed our neighborhood, sprinkling its invisible, stinking dampness over our yard and home.

If you have any hope at all when you are diagnosed with pancreatic cancer, you will have a particular kind of cancer in a particular part of your pancreas, making you a candidate for the Whipple procedure. If that's the case, a surgeon will carve out the upper third of your small intestine, your gall bladder, and most of your pancreas. The inventor of this magic act, Dr. Whipple, born in Iran in 1881, read Arabic, English, Farsi, French, Greek, Latin, and Turkish. (He had no need for pesky translations.)

If you are rich and well connected and excruciatingly lucky, you might finagle a liver transplant after your stage-four pancreatic cancer spreads to that organ. You can, for example, do what Steve Jobs did: hire someone to research all your options and learn that there is no waiting list for livers in Tennessee. Even if you go there and get a new liver, it's only a temporary solution.

Jean Fernel, the father of physiology, explained that the pancreas was the seat of melancholy and hypochondria. Pancreatic cancer cured my mother of hypochondria and melancholy. She had often complained of this or that, sore joints and headaches and nagging coughs. After she received her pancreatic death sentence, her complaints evaporated. She slept well for the first time in decades; her arthritis vanished; the bone spurs in her feet no longer hurt. Unlike many pancreatic cancer patients, she did not become depressed. "I feel fine," she said after being told her cancer was inoperable, and after being told there was nothing the doctors could do, and again after learning she had four to six months to live, and a final time two days before her death.

In a children's book about how we die, translating frightful ideas for fresh young minds, I found a photograph of a human pancreas cell. It is gorgeously ugly, hideously beautiful: crimson globes embedded in a pinkish-tan oval, all nestled on a bed of cabbage-olive green, spun through with gossamer gold.

SARAH K. LENZ

The Belly of Desire

1. Belly: the protruding abdomen of a pregnant woman

When I was five, I played the Pregnancy Game with the Hruby sisters, Jennifer and Stephanie. We used my Cabbage Patch Kid, Tallulah, a coveted Christmas present in 1986, but one that turned out to be a disappointment because she was so ugly. Tallulah's hair—short loops of yellow yarn that haloed her head—looked like a helmet made of a hook rug. I hated her for that. Still, I played with her often enough that her soft limbs developed a patina of grime.

After dinner, while our parents sat around the table talking, we girls went to my bedroom to play. We fought over who got to stick Tallulah up her shirt, but I won because she was my doll.

"I'm going to have a baby!" I cried. Then, as I had seen my mom do when she was pregnant with my little sister, I practiced Lamaze breathing—whooo, whooo, heeeeee.

Stephanie, playing Nurse, had me lie down on the bed so she could stick my arm with a fake hypodermic—a retractable Bic pen. To us babies came from bellies. We knew nothing of birth canals or uteruses, so every delivery necessitated a C-section. Doctor Jennifer reached in under my shirt with the scalpel—improvised from a plastic picnic knife with serrated edge—and swiped across the flesh of my abdomen, grasped Tallulah by her ankle, pulled her from under my Care Bears sweatshirt, and dangled her, triumphantly, above my head.

"You have a healthy baby girl," she said.

Once born, Tallulah was never fed, diapered, or rocked. My

desire for child rearing began and ended with the drama of birth. The game was about being the center of attention.

2. *Tummy: slang for paunch*

I got my tummy in the third grade. It came from eating too much fast food at the Chanticleer Drive In. That was the year Mom went back to work and stopped cooking, the year Dad lost the farm and we moved to a house in town. I was grieving my old home. As if by filling my tummy with french fries and hot fudge sundaes, I could fill a deeper emptiness.

Sometimes I sat naked on the toilet before bathing. I hunched over, grabbed the thick snake of fat that started at the side of my waist, pinched it away from my rib cage, and imagined ripping it off. When I reached the center of my abdomen, I had two handfuls of fatty flesh. My belly button disappeared into the fold. What remained looked like a giant, puckered mouth. I squeezed these blubber lips.

In the weeks leading up to my ninth birthday, I pored over Mom's Wilton-method cake decorating magazines, flipping through pictures of exquisite frosting designs. Pastel-green leaves unfurled over paper-smooth fondant from a cascade of buttercream-frosting roses, the petals so delicate they looked like real flowers. The children pictured next to the birthday cakes fascinated me too. They were thin and smiled freely. I, on the other hand, sucked in my tummy anytime I thought someone might be looking, especially the eye of a camera. In most photos I wore a tense, pained look on my face, more grimace than smile.

I settled on a chocolate sheet cake trimmed with a shell-frosting border, a cascade of pink-buttercream roses, and my name inscribed in pink frosting at the center. In the birthday snapshot, I wear a pink sweater stretched tight over my round tummy. I'm so caught up in the moment—seconds before I scoop one of those coveted roses off the cake and into my mouth—I forget to suck in my tummy. I smile, fat chipmunk cheeks framing my face. I look so happy knowing how the desired sugar will comfort me.

3. *Midriff: middle part of the body between chest and waist, often revealed to incite sexual appeal*

The summer after third grade, my cousin Erin came to stay with my family. Having abandoned the Pregnancy Game, the Hruby girls and I taught Erin our new favorite: Dirty Dancing. Inspired by the 1987 film starring Patrick Swayze (which I watched on VHS until I could repeat every line), we took turns dancing to the soundtrack. Still years from puberty, we couldn't see what was "dirty" about gyrating hips, but we got the love story. By the end of the film, Jennifer Grey's character, Baby Houseman, won the love of Johnny, the sexy dance instructor. It inspired us to emulate her every move.

For Christmas that year, my mother bought me the soundtrack, a vinyl LP. I played it on my Fisher-Price turntable. Using the song list on the album cover, we made elaborate dance lists, plotting out who was allowed to dance to each song.

My song was Eric Carmen's hit, "Hungry Eyes." In the scene at the dance studio, Baby wears a pink bra and matching pink shorts. Johnny dances in front of her, and Penny (the dancer Baby will sub for once she learns the routine) dances behind, her hands on Baby's back and waist, guiding. In this composition, Baby's midriff is the focal point, centered on the screen. Her chiseled abs undulate under the slight twisting movement each dance step triggers. Sweat glistens on her skin.

In classic '80s-style montage, the film cuts to a series of dance practices, all revealing that delicious midriff, while Carmen croons: "I've got this feeling that won't subside. . . . Now I've got you in my sights with these hungry eyes." In the last cut, Baby wears a tight white crop top. Then we see only her torso, the waistband of black Jockey panties, and Johnny's hands at the side of her waist.

With the recent development of my own midriff, Baby's image was one to aspire to. Surely when I was older and had breasts, my torso would look like that. Baby gave me something to hope for while miserable at school and scared about how much my parents worried about money. Though I didn't feel sexual attraction yet at that age, the way Johnny held, hugged, and laughed with Baby in the movie filled me with a feeling that I couldn't describe.

4. Jelly Belly: a gourmet jelly bean made in fifty-one flavors popular in America during the 1980s

When I was in fifth grade, every time Grandma B took me to the mall, she bought me jelly beans. Grandma had roughly the same measurements as Buddha. She stood about five feet one, measured forty-five inches at the waist, and possessed wide, child-bearing hips that allowed her to have nine babies. Even without all that childbirth, she would have been bottom heavy. All the women on that side of the family have inherited her shape.

I had tried on twenty pairs of jeans at JCPenney, but the only pair that fit around my belly had hideous front pleats and an ugly acid wash. Grandma insisted on buying not one, but two identical pairs of them since I couldn't get by with only one pair of school jeans. To my classmates, it'd look like I wore the same pair every day, making me seem even poorer than we were. As we stood at the Jelly Belly store counter, I tried to forget the shopping bag hanging from my wrist.

Though you could pick and choose separate flavors, I was too greedy, too worried about making the wrong choice and missing out. I loved the brilliant colors—cactus-pad green, stop-light red, buttery yellow, blue indigo, all luminous like seaglass. I wanted them all. So I ordered my usual—a half-pound assorted. The cashier weighed the candy and shoveled it from bin to white-paper bag with a silver-handled scoop.

When I clutched the bag, the beans clinked like pebbles, each a promise of pleasure. I cupped my eye over the opening so I saw a tunnel of white paper bag, and the end of it, a riot of bright-colored beans splendid as any kaleidoscope.

5. Muffin top: fatty flesh that spills over the waistline of pants or skirts because of tight clothing and/or excess body fat

When I was growing up, dressing up meant wearing No Nonsense Control-Top pantyhose, a modern replacement to the girdle, packaged in plastic orange envelopes. For church I wore dresses Mom had sewn from Simplicity patterns. On my body the finished garments never looked like the girls' on the pattern envelope. The bodices were too big, the waists too tight across my stomach.

"Control tops will fix that," Mom said when the fabric puckered unflatteringly over my gut.

She helped me pull the nylons up my legs, pinching my skin in an attempt to grasp the flimsy material. The waistband dug into my flesh, but as soon as I pulled my dress over my head, the waistband rolled down, making a sausage. I pressed hard on my lower belly. Swathed in tight elastic, it took on the tautness of a basketball while forming a crest of muffin-soft flesh above the waistband.

By the time I started college I'd stopped wearing control tops. They suffocated me. And I didn't need them anymore. The clothes off JCPenney racks fit me fine. Though I wouldn't have described myself as thin, I kept my weight down by giving up junk food and by exercising.

I wore control top panty hose only once after leaving home, when my Mom and I went to a cousin's wedding and shared a hotel room for the weekend. As we were dressing, I snagged a runner through my pantyhose.

"I have a brand-new pair you can have," Mom said. She rummaged through her luggage and handed me the orange package, unchanged over the years. As I shimmied the tight pantyhose up my legs, she kept up a steady commentary on my appearance.

"*That* lipstick's awfully bright for your complexion," she

noted as she fussed over her hair in the bathroom mirror. Her tone reminded me of Grandma B.

She nodded to my dress hanging from the towel rack and then to my Mary Jane flats. "Those shoes don't go with that hemline. Are they the only ones you brought?"

Standing there in my bra and pantyhose, I looked down at my belly. The pantyhose seam reminded me of a scar. I wanted to rip the nylons off my stomach. Instead I helped my mother zip her dress. Like wearing pantyhose, this wedding was something to be endured.

Since middle school, I had compared myself to the bride, Melissa, and always came up short. She was thinner, prettier, and richer than me. For years I'd wanted to shuck off the fat around my midsection, but standing in that hotel room, a new desire came over me. I wished I were the bride.

6. Six-pack: a set of well-developed rectus abdominis

My junior year of college, I dated a soldier. He had come back from Iraq hard and chiseled. As we lay in bed, I traced his six-pack with my tongue, then lightly kissed the six rippled muscles running from his pubic crest to his ribs. He was tan, lean, and smiled like a supermodel. My head told me, "He's sexy," but my body wasn't into it. I was so uncomfortable around him while naked, I could never climax. Once he ran his hands over my naked body, but stopped suddenly at my stomach.

"What's wrong?" I asked.

"You're just so—soft," he said. "Not fat, but soft. You've got to do some sit-ups or something." He grabbed at my belly and shook his hand like a dog wrestling a chew toy. I fought back tears when I thought about those hours of Pilates I had done when he was deployed, and still my body didn't make him—or me—happy.

When I got sick of faking orgasms, I broke up with the soldier. Even though I had done the dumping, I still felt like a failed husband-catcher.

7. Solar plexus: nerves of the sympathetic system located at the pit of the stomach

I found my solar plexus after breaking up with the soldier to date the man I would eventually marry. I had enrolled in a Hatha yoga class at the university fitness center. Chakras, I learned, were the body's energy centers. The solar plexus chakra controlled my fear, anxiety, personal power, and transitions.

"Let your belly soften. On the inhale let it grow large. Grow round," the yogi instructed. To nurture calmness, he explained, you must breathe from the bottom of your belly.

We did Mountain Pose—or *Tadasana*—which didn't look like any yoga poses I'd imagined. We stood. Feet aligned parallel, hip width apart. Eyes closed. Arms hung with palms out. Energy floated through my body and tingled out my fingertips.

"You're like seaweed floating in the ocean," the yogi said. "Your bones can float around in your body." He told me to place my hand on my belly.

"Your belly is an air pillow," the yogi said. "You know when you order something from Amazon, and it comes with those little plastic pillows full of air?" I chuckled. He was right. Just like an air pillow, the flesh cupped in my hand was somehow firm and soft at the same time. I imagined a radiating yellow ball of energy revolving just under my solar plexus. My belly didn't feel fat. It was strong. Whole.

When the session was over, I stood as tall as a mountain, proud of my rugged crags and jagged outcroppings.

8. Potbelly (1): a swollen or protuberant stomach

While dating the man I married, we took in midnight showings of cult-classic films in Midtown Omaha at the Dundee, a historic theater, the interior of which resembled a red velvet cake. Until I saw Quentin Tarantino's 1994 film *Pulp Fiction* there, I never thought a potbelly could be sexy. Fabienne (played by

Maria de Medeiros) obsesses over blueberry pie and potbellies. Her lover, Butch Coolidge—the boxer played by Bruce Willis—comes to her on the lam after killing his opponent in the ring.

She lolls on the bed of their dive-hotel room wearing a baggy T-shirt and a pair of cotton panties and says in a sexy French accent, "I've been *theenking* about potbellies. How I wish I had one."

"You should be happy, 'cause you do," Butch tells her.

She bristles, explaining that having a bit of tummy is not the same thing as a potbelly. Butch is perplexed and agitated, but she goes on.

"Potbellies make a man look either oafish or like a gorilla. But on a woman a potbelly is very sexy. The rest of you is normal. Normal face, normal legs, normal hips, normal ass, but with a big, perfectly round potbelly. If I had one, I'd wear a T-shirt two sizes too small to accentuate it."

"You think guys would find that attractive?" he asks.

"I don't give a damn what men find attractive. It's unfortunate what we find pleasing to the touch and pleasing to the eye is seldom the same."

Potbelly (2): having a round, protruding shape

If you Google the word "potbelly," one of the hits will take you to photographs of the Guatemalan potbelly sculptures. Hundreds of them dot the Pacific slope and highlands. Human figures portrayed as crude spheres with large, jutting guts. As if holding a ball, they cup their hands to their bellies. Some look fierce. Others tilt their heads back and stare into the sky with hollow, sunken eyes.

When I look at the Guatemalan potbelly sculptures in grainy black and white photos from archeological magazines, I imagine what it would be like to visit a sculpture, to walk along the wind-whipped coastal paradise, and to feel dwarfed by its twelve-ton body. Carved from basalt, a volcanic rock amalgam of large and small minerals, the sculpture's surface can be at once harsh and crude, smooth and fine. I want to run my hands over its rough velvet because it's one of the most beautiful things I've seen.

At first my attraction to these crude lumps puzzles me, until it dawns on me that until now my standards of beauty have had three sources: glossy magazines, the movies, and my mother. To all three I owe my shame at having a pronounced gut. For as long as I could remember, two contradictory and paradoxical desires—the wish for a flat midriff and the wish to be a mother—had tugged at me. I never questioned them, let alone their sources. Now I wondered if those urges were genuinely mine.

The pot belly statues were beautiful precisely because they shocked, and in doing so, broadened my sense of what is "acceptable." They're beautiful the way a Picasso painting is beautiful because it challenges staid notions of "beauty."

9. *Baby bump: the protruding abdomen of a visibly pregnant woman*

After I turned thirty and had been married for five years, photos of pregnant bellies started crowding my Facebook. One by one, women I'd lost touch with over the years—the Hruby sisters, my cousins—appeared on my Facebook page, their bellies bloated by pregnancy. They pressed bellies up to sinks for bathroom-mirror selfie shots. In professional portraits, they cupped their hands under their bellies and gazed lovingly down at their round protrusion.

When my younger sister got pregnant last year, it was the first time she ever beat me to a milestone. Putting careers first, my husband and I have delayed starting our own family. The intense jealousy I felt over my sister's pregnancy—when I wasn't even trying to get pregnant—startled me. The latent longing for a baby suddenly sprang to life. It was like wanting Baby Houseman's midriff all over again or envying Melissa when she was a bride (who, by the way, had just birthed her third beautiful baby).

The Christmas before my sister gave birth, she hired a photographer. She wanted to document her baby bump as it strained the elastic of her maternity jeans, and she wanted a family portrait too. One sunny day in early December I found myself with my mom and sister in a field with a photographer, who prompted us into posing.

"Put your hands on her belly," she told us. My mom and I followed instructions. I'd never touched a pregnant belly before. I was shocked by how hard her bump felt, smooth as a wind-polished boulder, yet under that stony hardness a delicate fetus curled, floating in amniotic fluid—a future member of the family already beloved by my sister and her husband.

10. Rectus abdominis: a long, flat, narrow muscle originating from the pubis to the seventh rib, forming part of the anterior abdominal wall

I was a year into a daily swimming and yoga regimen before I realized I had developed visible ab muscles. One evening I was packing for a business trip while my husband drank a beer and lounged on the bed watching me, hoping we'd have sex before I left. I tried on a new blouse, something slinky and tighter than anything else in my wardrobe.

"Does this blouse show too much belly?" I asked.

"You don't have belly to show."

I swiveled my hips in front of the mirror. He was right. For the first time in my life I didn't look like I had a fat belly. After years fighting my flabby stomach with exercise, I had won. It took me a while to realize this because I had long ago stopped exercising to shrink my waist measurement. Somewhere along the line it became its own pleasurable routine, satisfying and satiating.

I took the blouse off, folded and packed it in my suitcase. I grabbed my husband's beer off the nightstand and stole a swig. My beer drinking days are numbered. I hope to get pregnant soon, and my relationship with my belly will change again. The women who posted Facebook photos of their baby bumps now tell me all about their tiger stripes.

11. Tiger stripes: stretch marks and scarring patterns remaining on a belly after pregnancy

My cousin Erin was the first of my generation to have babies. She started when she was twenty, and now, almost fifteen years later, I still haven't tried to get pregnant, and she's still saving up for plastic surgery on her pregnancy-wrecked belly.

"Want to see something gross?" Erin said the last time I saw her. She lifted her shirt, revealing squishy, loose folds of skin that hung from her thin, petite frame between ribs and hips. The skin rippled with white scars.

My friend Jodi sees her tiger stripes differently. She told me about her four-year-old daughter's curiosity. "When Elizabeth asked me what those marks were on my belly, I said, those are my tiger stripes. They're special because I got them when you were in my belly. You'll get tiger stripes too someday after you have a baby." Jodi smiles when she tells me how excited Elizabeth is to get stretch marks someday. I wonder if Elizabeth plays the Pregnancy Game with her dolls.

I try to project into the future. Will I conceive? And if I do, how will pregnancy change my body and my relationship with it? Wanting is different than having. I won't know until I have them if tiger stripes will make me feel inferior or fierce. Still, naming what I want must count for something. I want someone, someday, to call me Mom.

The Population of Me

FOR MOST OF my life you've been the size of walnuts. Nuts and other produce are standard measurements for small bodily organs and glands. The prostate gland is also the size of a walnut, but by the time a man is forty, sometimes it's as big as an apricot. The pituitary gland is the size of a pea. Whereas we measure hail by sports balls. When hail gets to be the size of golf balls, that's when it starts to get serious. Someone's going to take a picture of that hail.

When I was born, stored inside the two of you, ovaries, was half the genetic material for a million people. That's about the population of San Jose, California. I've never been to San Jose, but I'm imagining it populated solely by those people, the potential humans whose blueprints you harbored. Tall men and short women. Dark-haired, with poor eyesight. If I were thinking of starting a business in that alternate San Jose, it would be an optical shop. There wouldn't be a lot of audience participation at concerts, as my people don't like to draw attention to themselves, but the shows would be well attended. The population would have to pay special attention to sunscreen. It would be a snarky San Jose.

But amid all those brown-eyed wallflowers, maybe there would be a few blonde-haired, blue-eyed standouts. They would be the product of long-forgotten recessive genes that give them the ability and inclination to be cheerleaders or to complete layups in a graceful fashion.

Those million people, though? They're only a fraction of the story. When I was a fetus, you held seven million egg follicles. That's the population of New York City, minus the Bronx. But to be brutally honest, most of those potential people didn't have a lot of, well, potential. So you performed a culling, keeping only the most promising million egg follicles. And of those million, throughout my lifetime, maybe five hundred will get even a shot at the big show.

I think of my San Jose with affection, all those ghost people, the artists, the bricklayers, the teachers, the petty thieves, the bureaucrats, the scientists. The odds were against them, were against all of us. But we made it. How can we ever look at another human being without a sense of wonder? Every person who takes a breath has scaled the Kilimanjaro of biology, has won the Powerball and is standing there with that big check and the shit-eating grin, or should be. We should be embracing each other as comrades, as survivors. Just showing up on Earth at all means we're winners.

I know we can't always get along. We need to argue about important things like the environment and the economy. We also need to argue about things that don't matter, like the Oxford comma and whether leggings are pants, because written into our blueprints are brains that want to make sense of things, that want to nail down the rules. Also written into our blueprints is the desire to have the last word.

But still, every so often I meet someone and I'm struck by the unlikelihood that I exist and that she exists and that we're in the same place, having a conversation, and we understand the words that the other says. And I feel a connection to her. When I read about a crime, I sometimes think about both the perpetrator and the victim and feel an almost unbearable sadness that there are perpetrators and victims, after all the work that was done behind the scenes, within the warm, dark factory of the human body, to bring them into the world. I think about my ghost San Jose, and all the other ghost San Joses, and about how we're the ones who made it into the outside world. We should be a little gentler with each other. We should be gentler with ourselves.

But back to you, ovaries. Of those million egg follicles you kept in safekeeping for me, two of them became living, breathing people. There's no mystery like that of a newborn baby. I never saw my job as molding my children, but more as letting their mysteries unfold. Deep in the genetic makeup of my son were instructions to curl his toes under his feet when he sits, like one of his uncles does. How far back does that go? Maybe five hundred thousand years ago there was a *Homo heidelbergensis* sitting on a rock in what would eventually be Germany, sharpening a stone point for a spear, with his toes curled under his feet.

Etched somewhere in the DNA of those two well-timed eggs were determination, a love of music, sharp wits, and a great capacity for kindness. Also (between the two of them, not naming names) stubbornness, perfectionism, a habit of telling jokes at the wrong time, and difficulty with time management. You did your part and I've tried to do my part. I tried to help them find their paths, but to also do a lot of staying out of their way so that they could become who they actually were and not who I imagined they might be. I didn't do much, but I think I didn't mess up your hard work. They're adult and almost adult, and so far, so good.

I'm grateful that these are the two I got, although I guess that I'd feel the same if I'd ended up with two of the ghost children in San Jose instead. I know I'm anthropomorphizing you way too much, ovaries, but you did a good job and I appreciate it.

Of course, you're only half the story. Testes are nothing to sneeze at either. They stay busy, cranking out genetic material on an as-needed basis, at an incomprehensible rate, like fifteen hundred sperm cells a second. It's survival of the fittest when it comes to sperm, and the culling is even more cutthroat than that of the egg follicles. In heat after heat in the race for life, almost all of them lose. Forget about my San Jose or even my New York City without the Bronx. We're talking twice the population of the earth every month. You, ovaries, are the students who prepared ahead of time. You read the syllabus and did the assignments the first week of class, and then just waited for the due dates. Testes are the students who goofed off all semester and then crammed the night of the test and still pulled off a pretty good grade. But don't resent the testes. It takes all kinds.

So, with all this talk of eggs and children, I want to point out that I'm much more than a mother. I'm a writer, I understand at least ninety percent of the rules of football, and I'm a highly competent parallel parker. Reproduction is only one of many facets of me. But you, ovaries, are unabashed in your single-mindedness. Everything you do is to advance the cause of extending my genetic legacy. I'm going to tell you something, but I think you know it already, if you're really honest with yourself: no more eggs will become people. For a while now, you've been acting as if nothing has changed, but really you've just been going through the motions, like a bookkeeper in an abandoned office. There's no new business, nobody's asking for the report, nobody's reading your emails, but you've still been keeping the books month after month.

At some point, though, your job will be done. Remember when you were the size of a walnut? Ligaments strained under the weight of the future generations within you. But eventually you'll be the size of an almond. I wonder what size nut you are now. A pistachio? Or maybe a cashew?

There's not too much plot to our story, and that's a good thing. You haven't been stricken with any diseases or major dysfunctions, although there's still time, I guess. Most likely one of my organs will eventually do me in, but sentimentally, I hope it's not you. You're my connection to the future and my link to the past. Just about every other part of my body could, theoretically, be swapped out for someone else's, or replaced by a machine. They just have their jobs to do. My heart pumps blood, my kidneys clean that blood, and my lungs supply that blood with oxygen, but they don't do their job differently from anyone else's heart, kidneys, or lungs. But you took pieces of me, of my mother, of my grandmother, of my great-grandfather, and offered them as possibilities to the future. You know the secrets of who I am and who I could have been. I'd like you to be with me until the end, like two old friends, side by side in rocking chairs, sitting quietly. They know each other so well that they don't have to say a word.

MATT ROBERTS

Pre-vasectomy Instruction No. 7

PRE-VASECTOMY INSTRUCTIONS[1]

1. Shave[2] genitalia up to the base of the penis; shower; and put on clean, jockey-type underwear just prior to coming to the office (see instructions[3] below).

2. If your appointment is first thing in the morning or first thing in the afternoon, we suggest that you do not eat that meal.[4]

3. Please inform your family/friends that NO OUTSIDE PARTIES ARE PERMITTED IN THE ROOM DURING THE PROCEDURE.[5]

4. Have someone drive you to and from the office. We suggest that you do NOT bring your children to our office to wait for you. There are sometimes delays and children can become anxious and excited. They have been known to want to jump on Daddy to greet him after surgery.

5. If you have a "normal" office job, arrange for at least a couple of days off from work after the procedure. You will be expected to "couch-potato" for two days.[6] If you do strenuous work, at least one week should be allotted for light activity only. Please discuss with the doctor in detail.[7]

6. PLEASE NOTE: The vas deferens is sent to a lab for confirmation.[8] You (if self-pay) or your insurance company will be billed for this examination/report.

7. You may want to bring along a walkman or CD player with headphones to listen to during the procedure.[9]

PRE-VASECTOMY RATIONALE

1. You can't afford any more children. You are finally ending each month with a little money in the account; you aren't buying gas and groceries on the credit card. This reasoning comes across as callous, selfish, insensitive, because it appears as if you are putting financial matters first, that you are a cold, calculating son-of-a-bitch.

2. You should instead start with how frustrated your wife seems lately, how desperate she is to find a job, to get into grad school. She has been a full-time mom for five years. She wants a house she can paint, a place of her own that she can refer to as home. You are both tired of living your life in other people's houses.

3. There really is no good time to have a baby, especially for you. It will be another year until your youngest is in preschool, then nine months to carry, another two years to be with the baby, and that's at least three years before she can even think to go back to grad school. Does she want to begin to forge her career now at thirty? Or five years later, at thirty-five? Your wife will counter with there is no good time to have a baby.

4. You know that there is no good time to have a baby. You learned that with Chloe[10] and then again with Carter. This is one of the reasons you offer: that it is time to start behaving responsibly. No more unplanned pregnancies. Your sketchy history with contraception leaves no other alternatives.

5. A few months ago, your wife kicked a hole in the wall. She was tired, frustrated, harried by the children, and she kicked a hole in the wall. Maybe at times she does the same thing the

father does: imagines for a moment a different life. It's perfectly natural. Nothing wrong with it, really, although we are often ashamed of ourselves for having thought it at all.

6. You look around at the walls of the house that you rent but do not own and find pictures of yourself as a child in a little red jacket alongside pictures of you as an adult holding your own child and know that nothing could be better than knowing that nothing is missing. And that's the question that the doctor doesn't ask: do you feel as if your family is complete?

POST VASECTOMY[11] INSTRUCTIONS

1. Upon arrival at home, sit in a comfortable chair, put your legs up and apply ice packs to the scrotum over your jockey-type underwear.[12] The ice should be in a plastic bag so as not to injure your skin. Keep ice packs on until ready to go to sleep for the night. Repeat the day after. This will markedly reduce the immediate post-operative pain and swelling.[13]

2. Use any pain medication supplied by your physician judiciously. Do not drink or drive if using narcotics for control of pain.

3. Keep incisions dry for two days. Keep Band-Aids on and replace as needed for 48 hours after surgery. Thereafter, personal preference will prevail. If bleeding or "oozing" occurs, apply pressure to the area.

4. Activities:
   • Sexual activity[14] may be resumed after 4 days if comfortable and no complications.
   • Do not tub bathe or go into a swimming pool for 7 to 10 days after surgery.
   • You may shower after 48 hours unless there is still some bleeding; if so, wait an additional 24 hours.

• Strenuous activity (biking, exercising, lawn-mowing) may be resumed after one week. Be sensible; if you feel any strain/pain in your scrotal area, stop the activity immediately and wait a few more days to resume the activity.

5. Contraceptive measures[15] must be used until two (2) negative sperm counts are obtained 2 and 4 months after surgery. If the second specimen contains sperm, further counts will be required.

6. A post-operative appointment will be scheduled one month after surgery. The incisions will be checked and the specimen bottles and lab instructions given to you.

7. Frequent ejaculations (average 16) will help to attain a negative sperm count sooner.

---

1 These are the instructions that I am given, that I go home with, and I pore over them alone in the bedroom sometimes, pondering what they are telling me that I must do. I can hear the doctor's voice in my head, his strange accent highlighting the pointing of a pencil at a laminated full-color diagram of the procedure itself. I don't catch much of the particulars because I'm thinking of where the cyst might be, but there is nothing in the paperwork about a cyst. I should have been paying more attention, because there is nothing in the paperwork at all about the procedure itself.

Later, my own research will tell me that a vasectomy is the surgical severing and sealing of the vasa deferentia, the two tubes that carry sperm from each testicle to the penis. After the procedure, the seminal fluid will not contain sperm, which means that I will be unable to get a woman pregnant. The most common form of vasectomy, the procedure that I will undergo, involves making an incision on one side of the scrotum, pulling the vas deferens through the incision, and placing a clamp on either side of a one-inch length of the tube. The surgeon then removes that section of the vas deferens, cauterizes the ends, folds each end over on itself and ties the folded section to the tube. Once the clamps are removed, the vas deferens is placed back inside of the scrotum and the incision is closed with stitches. The procedure is then repeated for the other testicle. An outpatient procedure, the surgery is performed using a local anesthetic injected directly into the scrotum and the whole process takes about thirty minutes.

According to the Mayo Clinic, about fifty million men worldwide have had a vasectomy with approximately five hudred thousand American men going under the knife each year. However, because a vasectomy often can't be reversed, the decision to undergo a vasectomy should be considered only when both partners agree that a permanent form of birth control is necessary; this may be because one partner has a health condition that makes pregnancy unsafe, or other birth control options have been exhausted for a variety of reasons, and, perhaps most importantly, no more children are desired. The finality of the decision suggests that a vasectomy should not be an option for those in an unstable relationship, those who are too young to be making such decisions about an unstable future, those who are counting on having children later by banking their sperm, or those who are unsure about whether or not their family is complete. Before making the decision, the Mayo Clinic suggests considering such things as what would happen if your financial situation improved so that you can afford another child; what if you and your partner separated, divorced, or one of you died; and how you would feel if one of your children died. "It's not uncommon for men to consider vasectomy in response to a stressful situation such as an illness, financial setback, death in the family or birth of a child. Because the stress may be temporary, however, it's better to seek counseling or psychotherapy to help you cope than to undergo a surgical procedure you may later regret."

A vasectomy means no more babies. Carter is turning two this weekend; Chloe goes to kindergarten in the fall. By the time my wife's parents return to Arizona or we ever get down to Louisiana for a visit, there won't be any more babies. We'll have kids. A little girl who is starting to read, and a little boy who doesn't wear diapers anymore. No more babies.

2 The morning of the Mardi Gras party, I run by Great Clips to have my head shaved. The girl has full sleeves and a blue star on her neck. She tells me she has been getting tattooed since she was fourteen. As she clippers my hair she has to step away for a moment, and my blurry

head is two toned in the mirror, my thick black eyeglasses sitting on the counter. They always ask me several times to confirm that I want my head shaved.

"Yes," I always have to tell them, "I'm going to Bic it when I get home." Only today, sitting there with my blurry head half-shaved, I want to talk about something other than the Mardi Gras party, something other than tattoos. I want to talk about why I am shaving my head again. I want to talk about my vasectomy.

In a hot shower at home I fill my hand with lather, consider it there in my palm for a moment, and then rub it across my head. I've already carefully removed the yellow cover from the head of the razor while I was outside the shower. The single silver blade is housed in a stiff white frame. No swivels. No lubricating strip. The quarter-inch of hair left on my head from the clippers clogs the blade with each short scrape. I often shave my head, but I can't ignore why I'm doing it now.

I will shave my head every other day for the next two weeks. I will cut myself today, sucking air in over my teeth when I do so. My skin will burn and peel away as it dries. If I miss a day the hair will grow longer and thicker and I will begin to cut myself again. It takes discipline. I must get better at it each time. Faster. Surer. Steadier-handed. I will not let my life be ruled by Fear and Doubt. I will not fear the razor. I will not bleed. I will not bleed. I will not.

3 SHAVING INSTRUCTION FOR VASECTOMY

Before the vasectomy, you should shave the pubic hair from around the scrotum as shown in this illustration. You can do it yourself or ask your wife or friend to do it for you at home; preferably the same day of the procedure. Do not use an electric razor. Use a safety razor, even a lady's razor, which is less likely to cause bleeding than an electric razor. Some gentlemen find that using baby oil instead of shaving cream is also more comfortable. Failure to shave at home will definitely delay the start of the procedure.

Beneath these words is the ghostly vision of an uncircumcised penis being nudged aside by some disembodied thumb to reveal a set of vein-covered testicles dangling perfectly. The image is obviously a bad black-and-white photocopy of a color image from a book or a pamphlet, and no one has bothered to change the toner cartridge on the copier for a while. A hand-drawn bracket is alongside the drawing with the words *shave entire area* in the delicate loops of the female office assistant's handwriting. The bracket line stretches from the top of the balls to the dark underside sagging there on the page. An arrow points to the top. The same handwriting as before tells the reader to *start here.*

4 Conventional wisdom suggests that eating shortly before surgery is a bad idea. I recall the kicked-in-the-nuts-want-to-vomit feeling from the initial consultation and the only-more-kicked-in-the-nuts-want-to-vomit feeling from my scrotal ultrasound and start washing some potatoes in the sink as some bacon browns in the cast iron fry pan on the stove. I move my hands across the surface of the potato, feeling the divots, the bumps, the topography of the thick dimpled skin of the thing as water pours over it, the water pouring over my hands. The peeler is heavy, a weight in my hand, and the blade drags across the skin, peeling away a shank of starchy white flesh. The skin falls away, slapping into the stainless steel sink. I turn the potato, pulling the peeler in toward the heel of my hand with each turn, this plain brown everyday thing in my hand transforming into something angular, something different. The thing slides around in my clenched fist, trying to slip out and bang around in the steel sink, but I won't let go. I don't slip. Hold tight. Little beads of starch punctuate the thing, glistening.

5 Jenn and I aren't talking about the vasectomy; we only discuss it as if it is a cancer or a genetic disease. Perhaps it is the finality of it all that lends this analogy some weight. Maybe it's this medical procedure's finality that has me so uncomfortable, so quiet, so reticent to speak of it at all with Jenn. I avoid it at all costs. Just do it. Let it become a memory. Something that comes up in conversation with a young man in the gym years from now. Maybe I'll say, "Oh, I had one of those," like it was baba ganoush or crawfish bread, a band seen at Tipitina's before they had a hit single. Or maybe it's the kind of thing that doesn't get spoken of at all, except for rare occasions of peculiar vulnerability when something just has to be said; the ghost of that finality lingering over my shoulder, making me question, even if only

momentarily but also sometimes every day forever, that decision that cannot be changed once it has been made. Regret is sometimes an ugly rumpled thing left on the doorstep, on the dresser, on the shelf above the washing machine, moved from room to room by bat-winged, red-eyed Fear and Doubt. I swore long ago that I would not let my life be ruled by Fear and Doubt, that my life would not be one of Regret.

6 A week after the Mardi Gras party, the neighbors from across the street have a party for their daughter, Grace, turning one year old today. At the party, while watching presents being unwrapped, I hear Carrie's sister-in-law ask Jenn, "How old are you?"

"He's thirty-five. I'm only thirty," Jenn says.

"Oh, you've got plenty of time. You shouldn't be getting snipped. You're too young."

We are too young. But we had ours early. We don't own a house. We have almost no savings, cashing out my previous 401K early to reduce our debt load. Earlier in the week I saw a dent and scrape on the passenger-side rear quarter-panel of the car we drive every day but do not own. We bought that car with a down payment from Jenn's parents. "Well, since we didn't have to pay for a wedding," her mother had said in the lobby of the dealership.

Chloe was only six months old when we bought that car. Winter was coming and the car had all-wheel drive and anti-lock brakes. We weren't ready for Carter either. We hadn't planned on Jenn staying home that long; we hadn't planned on another child when we bought that car. We were filling the tank of that car with the credit card so we could drive it to the store to buy groceries with the credit card. I was staying up late grading papers after coming home from my second job. I would sometimes go an entire day without seeing my children; Chloe's tiny body sometimes slumped on the sofa because she tried to wait up for me.

But the sister-in-law doesn't see any of that. The sister-in-law doesn't see a young woman with a college degree who has been changing diapers for almost five straight years, who wants so badly to sleep through the night just once, who was first in her family to graduate from college—crossing the stage five months pregnant with her first child. The sister-in-law doesn't see a woman who wants a career of her own—the career she desires instead of settling for a job that gives her something to do while the kids are at school. She doesn't see Jenn crying with a baby in her belly wondering *what are we going to do?* She doesn't see Jenn going to and from the hospital every two hours to express her milk into bags for a syringe that will be taped to the back of a chair so gravity can pull my wife's breast milk down through the GI tube entering our baby's left nostril and down into the stomach. Twenty-eight days she was in the hospital. Four pounds, two ounces. Two months early.

But the sister-in-law doesn't see that. The sister-in-law doesn't see the doctor from Tulane pull my son from my wife's belly with a large white suction cup because the umbilical cord had wrapped around his neck, the hematoma blossoming on his head as a lump, a thick purple bud of soft new head.

The sister-in-law doesn't see the letters of rejection from grad schools. The way Jenn started talking about giving up on school all together, settling for something else for the rest of her life. The sister-in-law doesn't see her have to turn down a job on the Arizona Governor's Drought Task Force because she can't quit the part-time job she has now until May. Jenn is devoted to the kids at school, an inner-city elementary in South Phoenix where she teaches an after-school science club for the service learning program at Arizona State. The sister-in-law isn't there when the letter arrived two days ago from Louisiana State University. Jenn has found a young professor willing to be her adviser. Jenn is ready to start pursuing her own dreams for a change.

"You're too young," the sister-in-law says again.

That evening, after the sun goes down and the men are left outside to drink beer, I chat with the neighbors who had vasectomies themselves only a few months ago. We are all talking about the shaving of the balls. Ben tells us about the lack of anesthetic and how the nurse critiqued his wife's shave job. He distinctly remembers her using the words "totem pole." Jimmy remembers when they attached the clamp, but the doctor shaved his balls. I look at Jimmy standing there with his can of Natural Light in a foam beer koozie and wonder how one man tells another man that another man shaved his balls, that he just lay there and let another person shave his balls from behind an unfolded paper gown taped to two lamps, the crisp, hygienic folds forming a grid that forced him to focus on the speaker above his head, although music is coming from a small boom box somewhere in the room behind the crisp, hygienic folds, but he can't see where it is, so he doesn't know where exactly the music is coming from—in fact, he can't be sure that the music isn't coming in from the speaker above him because the Valium kicked in while he was in the waiting room— and he remembers listing toward the wall as he passed a fat doctor in front of the vending machine, and then I hear Jimmy say, "Valium?"

"Hell, yeah, you should fill that prescription," he says, fingering the bill of his ball cap. "Take one of those, chase it down with a beer or two. You'll be fine."

7 The story of my vasectomy starts two weeks ago with a visit to the urologist's office, when I am standing in the examination room with my hands on my hips and my jeans around my ankles as a grown man fondles my nuts. He is nice enough, maybe a little too professional, but I imagine that when you spend a good part of your day inspecting grown men's testicles, you don't want to be too friendly.

He is wearing navy slacks, a blue pinstripe oxford shirt, and rim-less glasses. His short hair is what I would describe as a "little boy cut" and he has a slight accent, although I can't place it. I tell him my age, how many children we have, my wife's age. His questions are not probing, and he quickly cuts to the point. Before he started examining my balls, he was asking me some questions:

"So you are finished?"

I tell him that my wife is eager to start a career. That she has applied to grad school. He is sympathetic ("And you don't want to get pregnant *then*"). I explain how we've tried other forms of birth control, but that things just haven't worked out: the bad experience with condoms, the apprehension about hormone therapy, the vulgarity of spermicide, the flushing of the IUD. But I want to talk about more than contraception. I want to talk about the salary of an adjunct instructor in the English department of an enormous university. I want to talk about how Chloe was born two months early; Carter, with his cord wrapped around his neck. I want to talk about how my wife has few friends out here in Arizona, how she wants a job, a career. How she wasn't living at home when her maw-maw died, how her parents are coming out here during Mardi Gras to get away from the devastation wrought by Hurricane Katrina. I want to tell him about the five straight years of changing diapers, about the twenty-eight days Chloe spent in the hospital, about our hopes for the next three years. I want to talk about anxiety, about dread, about fear and doubt. I want this to be about the future. I want someone to listen to me, listen to the logic of my having come to this choice, to this juncture, to standing in a room with my pants around my ankles while a strange man with an unidentifiable accent tugs gently on my balls.

8 The urologist's hands move back to what he referred to as a cyst. He is talking, referring to it using medical terminology that I don't remember because the words *there's a cyst on this one* keep repeating over and over in my head. He recommends an ultrasound, his gloved hand still cupping my balls.

"This is nothing to worry about," he says, perhaps detecting a rise in my pulse rate through the blue veins on my balls. "I've seen hundreds of these. If it was cancer, then I'd be able to tell. This is nothing."

"Good, because you're really freaking me out," I say.

"This is nothing," he repeats. But he requests an ultrasound anyway, suggesting that he simply wants to know where the cyst is exactly. I leave the office with a receipt, a copy of the privacy policy waiver, a prescription for 10 mg of Valium, a set of instructions, and the feeling that I've been kicked in the nuts. I will later learn that the benign lump lurking in my balls is called a spermatocele. According to the Mayo Clinic, "a spermatocele is a cyst that develops in the epididymis—the small, coiled tube situated in the upper testicle that collects and transports sperm. Generally painless and noncancerous (benign), a spermatocele usually is filled with milky or clear fluid that may contain dead sperm. Spermatoceles are usually less than two to three centimeters in diameter, although some may be larger. Most of the volume of the cyst is fluid. A common condition, a spermatocele doesn't impair fertility, nor does it require treatment unless it grows large enough to cause discomfort."

My research will tell me that the difference between a tumor and a cyst is academic. A cyst is a fluid-filled sac while a tumor is a swelling. Tumors are synonymous with cancer, but even a large bruise, a hematoma, is a tumor. Not all tumors are cancerous, nor are all cysts. Some cysts, however, like some tumors, can prove to be malignant. The only way to be sure if the cyst or tumor is benign, noncancerous, is to perform a biopsy, removing some of the tissue for examination under a microscope.

9 On the night before my ultrasound, walking from the living room into the kitchen, I imagine myself on the operating table. I want to say that I have the Pixies' "Wave of Mutilation" playing in my head when I imagine this scene, but that would be dishonest of me. I am not thinking of the Pixies' "Wave of Mutilation." The music in my head when I have this fantasy is a Beatles song. I don't really know any Beatles songs by name. I know things like Yoko broke up the band and that there was a different drummer before Ringo; that Paul and John apparently didn't get along that well; and as all of us can, I can recognize a Beatles tune as a Beatles tune in instrumental form when using downtown elevators. I can hear the music in my head when I think of this ending, the words repeating over and over and over in my mind: *Woke up, fell out of bed, dragged a comb across my head.* I want to say the Pixies because I hate the Beatles.

10 Chloe was born two months early. Four pounds, two ounces. We consider ourselves lucky. She was off the ventilator in less than twenty-four hours and then off oxygen after only a week. She remained in the hospital for twenty-eight days and had to learn the normally instinctual skill of feeding. A small tube entered her nose and ran down her throat into her stomach, and it was through this tube that gravity pulled my wife's milk from a syringe taped to the back of a chair.

At the end of those twenty-eight days, we brought Chloe home, a day before my birthday, to my mother, Chloe's grandmother, visiting from Louisiana. My mother had been worried. Her own only little girl died three days after birth, un-baptized and un-named. One afternoon, my brother Ricky sat with me in the hallway of my parents' house showing me pictures from his med school textbooks, showing me pictures of babies whose brains and spinal columns had formed outside the fetus's body. Not exactly this, he said, but something like this. I saw the little green marble slab at the family plot on a cold November day in Chicago when we buried my mother's father, my grandfather. I remember riding in the car one day with Mom after I had graduated from college when she said, "You know, you might not be able to have a girl. Roberts boys might not be able to have girls." So my mother worried. I worried. That afternoon at the hospital when my daughter was born, when the nurses started hurrying up and calls were being made about rooms and who's on call, the nurses kept telling me, "It's a girl. Keep your fingers crossed for a girl. Girl preemies fare better than boys. They develop faster in the womb." I didn't know what to hope for; when they asked me to cut the cord, my hands were numb. I stood there and stuttered, lightheaded, tiny lights flashing before my eyes, holding up my hands as if they were made out of rubber. I was soon shoved out of the way, and our baby girl was wheeled out of the room under a plastic dome.

We were lucky. Our child's life would be free of visits from occupational therapists and debates about mainstreaming. Nobody had to make their peace around a tiny white casket, another tiny plaque set in the ground. The day my mother arrived, she didn't know that Chloe was home with us. She was talking to Jenn with her back to me, and was still talking when she turned around to face me, not even noticing the tiny bundle I held in my arms as she talked hurriedly about going to the hospital to see the new baby. I passed Chloe to my mother, and she held those five pounds of baby Roberts girl and wouldn't let go until she left for the airport days later.

Almost two years later Chloe would still be small, barely registering in the lower percentages on the charts that tracked her progress, but she weighed nineteen pounds and stood thirty-one inches tall. Despite a brief period of physical therapy, our preemie child avoided major health problems. She caught up quickly with a prescription for Zantac, and I would feel pangs of guilt anytime I saw the parents of horribly afflicted children standing alongside some torturous wheelchair contraption that helped these kids to assume a hideous approximation of standing erect. I'm not sure how they do it, what reserves of strength these people tap into, but it is truly humbling for any parent who has kicked a hole in the wall during a forty-minute crying jag over a piece of candy to witness such extraordinary acts of devotion. Today, my daughter is an ordinary kid. She uses the potty on her own but still has accidents; we sometimes have to groom the twisted dreads out of her hair with apple-scented detangler; she dresses herself in orange and black striped pants, pink ballet slippers, a red long-sleeved shirt with a rhinestone heart, and a rainbow hat; she can put her own shoes on (but can't tie laces), helps make the pancakes on Sunday morning, counts to twelve in Spanish, and dreams of one day having a little sister.

11   I am blissed out on Valium while lying on a padded examination table with my torso behind a blue curtain taped to a lamp as the urologist sings along to Supertramp's "The Logical Song." But right now it is Bob Seger's "Rock and Roll Never Forgets," then U2's "Pride (In the Name of Love)." A small boom box in the corner of the room is tuned to the radio, rolling out an endless stream of classic rock and station identifications. Tiny clinking noises mix with muffled voices as the urologist and his assistant whisper to one another about any number of things. My brain smiles for a moment before realizing that it's listening to the Bee Gees' "Jive Talkin'" before quickly giving way to Aerosmith. My vasectomy has a classic rock soundtrack. Sitting behind that blue sterile curtain, I can see my life flash before my eyes as I stare at the useless speaker on the ceiling and listen to the rhythmic chugging of Joe Perry's guitar work on "Sweet Emotion." I can hear my father's Buddy Holly records playing on the turntable in the living room on Saturday afternoons. He loves the Rolling Stones ("greatest rock and roll band in the world") but despised my older brother Ricky's fetish for Led Zeppelin ("just a bunch of noise"). Dad was a nylon-jacket-wearing member of the Shades of the '50s classic car club, and Ricky was in a Led Zeppelin cover band that rehearsed in our garage. They cranked out "Whole Lotta Love" in Archbishop Rummell's gymnasium for the talent contest, Ricky's Gibson Les Paul glowing gold in the spotlight, the gold of trophies almost as tall as me in rooms around the house. Tom had a beer can collection and *Kiss Alive II.* Soon there were too many cans to keep, and Tom would bring me with him to the Mississippi Gulf Coast Coliseum to see AC/DC. "Dirty Deeds Done Dirt Cheap" was menacing the kids around the pool at the country club; I stopped wearing my Speedo when my body discovered Laura H—'s body. Laura tore Tom's Iron Maiden and

Kiss posters off the wall one day when I had to lock her in until family and friends could collect her. She was the daughter of a psychologist and was obsessed with Pat Benatar. Before her psychotic break, Laura had invited me to her house to play spin the bottle. I guess I lost because I left the house that day having never been kissed. My dashboard pushes out Allman Brothers and Canned Heat as I leave New Orleans at twenty-six, still a virgin and having never seen snow, to live in Colorado with Jenn. Our babies were baptized in the blues, me crooning Taj Mahal's "Johnny Too Bad" to Chloe as I held her in the Special Care Nursery; Carter's middlenamesake the uptown bar on the corner of St. Charles and S. Carrollton where Jenn and I had our first date listening to John Mooney play slide guitar with a beer bottle. There were lots of other bands, lots of other music, and as today, what constitutes classic rock will one day change. Someday I will turn on the radio to a classic rock station and hear songs that I don't recognize, but my oldest daughter, behind the wheel with a learner's permit, will. I probably won't like it, probably call it just a bunch of noise, and I will understand how my father's generation must have felt about my generation's lack of regard for the Beatles. I lie on my back in the urologist's office smiling at the muffled clinking and tiny voices, the blue pool of my youth as I came into puberty glowing in the back of my head beneath the neon sign from the top of the WRNO building. I smile because I know that things will be different. I can see the future. I know that this family is complete. And as that future comes into scalpel-sharp focus, Paul McCartney's voice rises over a succession of earnest piano notes: *Speaking words of wisdom, let it be, let it be . . . There will be an answer, let it be, let it be. Let it be, let it be . . .*

12 I'm standing in the bathroom with a large dollop of Barbasol cupped in my hand beneath my balls, hesitating for a moment before I lift my hand to smear them with the shaving cream, looking at the grizzled hair frizzed outward from the wrinkled skin. It's cold in the bathroom and my balls have shrunk in toward the warmth of my body. This is obviously going to be harder than I expected.

I hold the plastic Bic single-blade razor up to examine the blade. No nicks. No scratches. I hold the plastic Bic single-blade razor up against my balls and pull down. No nicks. No scratches. No shave.

The shaving cream is too thick, the blade simply sliding across the surface of the skin as the hair is matted down.

I get in the shower, get it good and hot, and prepare to shave without any kind of lubricant, hoping to soon have a clean, hairless sack resembling a plucked chicken. As per the illustration accompanying the instructions, I use my thumb to hold the rest of me out of the way, start at the base, and proceed to *shave entire area*. The thrum of the water against my back echoes the mantra playing over and over and over in my head: I will not bleed. I will not bleed. I will not.

13 Surgically induced sterility does not affect virility. Because the testicles continue to produce the same amount of the male sex hormone testosterone after a vasectomy, the procedure should have no effect on sex drive or the ability to get an erection. A vasectomy does not affect body composition and other male characteristics. Men will still have the same muscle mass, facial hair, and voice. A vasectomy should not noticeably reduce the volume of ejaculate, since semen consists almost entirely of fluids produced by the prostate and other glands. I find this last point pleasing for some strange reason, although I am disturbed to learn that, as with any invasive surgical procedure, vasectomy may cause some complications, such as bleeding within the scrotum, infection at the site of incision, severe scrotal pain that lasts up to three months, and the formation of a sperm granuloma, an inflammatory mass caused by the leakage of sperm into the scrotum. I have a cyst; I don't want to even ponder sperm granuloma but find myself repeating the words over and over again in my head. Sperm granuloma. Sperm granuloma. Sperm granuloma.

14 My friends drool over the idea of unlimited unprotected sex. But it's not about sex. It's about anxiety. It's about uncertainty. It's about dread. It's about Fear and Doubt, the dark, curled horns that protrude from the forehead of gray-skinned Regret. He is always there, perched somewhere with his chin in his hand, watching you. I have to admit, though, that sometimes a thought sparks, catches fire there for a moment. I'll lie there in bed alone, rubbing against the sheets in a half-slumber, while Jenn is crowded under the kids in our bedroom. That spark is most often doused in the shower. I was a virgin until I

was twenty-six years old, but I've been masturbating since I was five. I guess I just didn't expect to be masturbating this much at thirty-five.

15 Jenn offers to see the doctor about an IUD because this procedure has me so freaked out, so worked up. However, today is Friday, and she is calling me at work because she can't get the kids dressed for a photo at Wal-Mart. My mother has been pressuring us to take a picture of Carter, now over two years old, in the little red jacket that all three of us boys wore, the framed photos hanging on the wall in the hallway of my mother's empty nest near pictures of my brother's own two boys wearing that very same jacket. Today Jenn is at the end of her rope, unable to get Carter's diaper on. Jenn starts to cry.

Carter had done the same thing to me earlier in the week. I was holding him down to get the pants on, pinning him to the floor with my hand flat on his chest, him screaming, full of red-faced fear and anger, his voice nothing more than a shrill rasp as tears poured out of his eyes, and I had to stand up and walk away, shutting him in Chloe's room behind the closed door. Carter screamed and pounded on the door, the brassy rattle of the doorknob shaking into our ears as Chloe and I sat on the edge of the bed they all shared with their mother at night. Chloe asked me why we don't just go open the door. I stared at my hands and had to explain to her that I needed to be away from Carter for a few minutes, that the only way to get away from him was to shut the door. Chloe leaned into me, wrapping her arms around me, as if to remind me that I am a good father.

"Shhhhhhhh," I say into the receiver softly, "shhhhhhh, baby. It'll be all right. None of that matters." I try to absorb my wife's fears through the phone line, let her fears become mine because I know what she fears. It's the finality of the act, the gray-eyed specter of Regret lingering around the house as our beautiful children grow older, their tiny red jackets tearing apart at the seams as dust is already collecting on the frames in the suddenly quiet hall.

AMY BUTCHER

Taking Shape

IN YOGA ON Sunday evenings, my classmates and I spread our mats over hardwood floors. We remove our shoes and then our socks and step lightly onto our mats.

Before beginning, the instructor reminds us that this is not a competition.

"Only with yourself," she clarifies, and then she dims the lights and turns on the rhythmic chanting.

I feel my body bend and shape, I feel it warm, I feel my young skin and my young bones. But halfway through, I feel the woman on the mat beside me watching my pose. Like others in the room, she is older than me, and she squints as our bodies take shape, wondering why my body can't shape better.

"On the next inhale," the instructor says from the front of the room, "I want you to push deeper, push farther."

I lean forward. I feel a grating in my right knee like sandpaper on bone. The woman on the mat beside me cranes her neck away, but her eyes are still on me. She notices my face. She notices the way my mouth curls.

The instructor pads around, inspecting our poses. I see her bare toes beside me, and then she places her hand on my abdomen, pushing upward. "Bend the spine," she says. "Lean into the knee. Stretch *deeper*."

I feel my face flush. My toes curl against the mat. I want to tell her that I can't, that it hurts, that the skin is only twenty-two but the joint feels older, worn away, rusty. I want to tell the woman beside me to stop looking, stop staring.

Instead I heave, grin, force myself farther.

In my earliest memory, I am three and in pain. My mother has driven me across town in the pouring rain to a dark room with hardwood floors because I move differently from other children. My steps are slow, sluggish. My feet slide across the floor in a slinked heave. I climb steps sideways, lifting my foot out and then up.

The woman we meet has thick red curls. When I take her hand, she leads me to a gel-filled mat in the center of the room. Suspended between two thin sheets of plastic are dozens of tiny toy fish, all them wobbling in neon pink gel. They move as I move. I sit on top of one striped orange and yellow. The woman kneels beside me and helps me untie my shoes.

"Oh, look at that," she says. She peels back white lace to expose my painted toenails. She feels for the other foot. "Are the nails on this foot pink, too?"

When my feet are bare, she rolls my pants up to my thighs. "We're just going to try some simple exercises," she says. "Can you wiggle your toes for me?"

The gel is cold against my bare legs and bare feet, but I listen to the woman. I do what it is she wants me to do.

"Just pretend you're pedaling an invisible bicycle!" she says.

I lean back on the mat and rotate my legs.

"And kick?"

I kick.

She asks me to bend my knees. I try, but it hurts. She asks me to stand and walk toward her. She holds out her arms as if I were learning to walk for the first time. But it hurts and I am young; I can't find the words to tell her that I've made up my own way to walk because of the pain. I shift my weight, as though the fish beneath me are real. I wobble.

"Okay," she says. "Let's sit back down."

I take my seat, and then she asks me if I know that I am different.

"Do you *feel* different?" she asks.

I don't say anything. Beneath me, the fish wobble with my breathing. Their movement is fluid, flawless.

"Hmm," she says. She touches my knees and explains that the place where most three-year-olds' legs connect is not swollen like mine. My legs are round in the middle, "like a snake that's swallowed a mouse," she explains, but I don't need her example to understand. There is a thick bulge in the center of each leg—two enlarged kneecaps, round as baseballs.

"Is that why you walk like that?" she tries again, touching the joint. "Do you walk like that because it hurts?"

I shift against the mat.

"Let's try it again," she says, as if this time, something will be different. I stand up. I fall back.

"Again," she says.

Weeks later, I wake to legs that won't bend. I scream until my mother comes. I tell her the truth: that I can't get up.

"Oh, for Pete's sake," she says, pulling at the drawers of my bureau. "Come on now."

The day before, my mother and I saw my neighbor in the grocery store on crutches—he had broken his leg jumping from a tree. My mother thinks I am imitating him, or that I want attention. She thinks I have gotten carried away with the idea of injury.

"No, really," I say. When she still doesn't turn, I pound my fists into pillows. "It hurts," I howl. "It hurts."

When at last my mother peels back the comforter, she gasps. She believes me. My right kneecap is larger than ever before, swollen and inflamed and red. She calls the pediatrician, cradling the phone against her shoulder as she carries me to the tub.

"Twenty minutes," my mother repeats to herself and then to me. She brings me a Pop-Tart on a plate so that I won't be late for morning nursery school. I watch the rainbow sprinkles fall into the bathwater and dissolve as I chew.

When the timer goes off, my mother nods, and together we watch as I rise from the tub and bend my leg slowly. It hurts less. The swelling has decreased.

"Good," she says. "Good." She towels me off tenderly, carefully, and helps me get dressed for school.

But within a month, I wake every morning to legs that won't bend. After a week, my mother buckles me into the backseat of her forest green Ford Aerostar. We drive forty minutes south to the Children's Hospital of Philadelphia. We park in an underground garage that smells like dirt and windshield wiper fluid, then take a glass elevator to the twelfth floor. My mother lets me push the buttons as Philadelphia comes into view.

"There's the Wachovia Building," she says, then, "there's Philadelphia City Hall. There's One Liberty Place." I listen to the names. I mouth them silently as the elevator climbs.

Inside the hospital, we meet with a man with a foreign name. Soon we meet with him every other week.

At the start of each appointment, he hands me a piece of gum, and I chew quietly while he talks to my parents in long, multisyllabic words I can't understand. The words sound foreign, and I like to listen to him, to his thick accent, and think of him as a king from a distant land. I chew the gum and imagine a throne surrounded by crowns and pits of shimmering jewels. I imagine golden canes. He squeezes my bare knees and smiles.

During one appointment, he tells my mother that she should withhold dairy. "Just to see," he says.

On the way home, my mother and I stop at a grocery store. While she locates the soymilk, she leaves me in the cereal aisle to pick any box that I want.

"Any kind," she says. "It'll be all yours."

I walk up and down the aisle slowly, carefully, touching the colorful boxes and cartoon faces. When I come home with Lucky Charms, my mother puts the box on the top of the fridge.

"So your brothers can't get at it," she says.

The next morning when Wesley whines, my mother shakes her head. "It's to mask the taste of her new milk," she says.

I stick out my tongue, grateful that the seizing pain in my knees has afforded me this much.

But the pain continues, even without ice cream, string cheese, milk. After a month, we return to the hospital, and a nurse in pink scrubs helps me climb into a paper gown. She takes my hand and leads me to a room full of windows. In the center of the room is a chair, and when I sit down like she instructs, she places Velcro straps around my wrists and ankles. She steps back to examine her work. I can see my parents standing behind her beyond the thick glass. With gloved hands, the nurse pulls at the seam in the rear of my gown. She peels the corners to expose my back. Then she takes a yellow plastic board down from a cabinet and places needles in rounded slots.

"This won't hurt," she says, lining the board along my backbone, but we both know it will, and when the needles—all fifty of them—break through my skin, I scream, I cry, I watch my parents watching me from behind the glass.

At home that night, I sit again in a tub of warm water. My mother cries softly beside me on the cold blue tile of the bathroom. She runs a wet washcloth down my back, moving softly, moving lightly.

"We thought you might be allergic," she says. "But you're not."

After my father towels me dry, he puts me on his knee and tries to explain. "Juvenile rheumatoid arthritis," he says, touching the joints, but it sounds like the doctor's words: foreign, unfamiliar.

That night, my parents snap on two pink splints that extend from my pelvis to each ankle. The splints realign the kneecaps, they explain. They prevent morning stiffness. After they tuck me in, they lie along the edges of my bed and read to me from *The Little Engine That Could*. I do my best to lie still, to listen to their words, to learn what it is to be straight like a rail.

"I think I can, I think I can," my father reads, and I imagine my kneecaps as the place where two rails meet—how it must be strong and straight, how it must align my bones so the train doesn't wobble off the tracks.

But after the book is done and they leave, I can't roll over or bend my knees or curl my body upward in an effort to warm my toes. In the middle of the night, I wake panting, hot, my legs on fire with an itch I can't reach. I peel back the comforter and try to squeeze my fingers between the plastic and my flesh, but the bulk of the splint pinches my fingers.

In the morning, I tell my mother about the itch. I tell her that I don't like the splints, and that they smell bad—like sweat, like body. My mother nods sympathetically. "We're doing all we can," she says. "This is new for us too."

That night, I drink a tall glass of water with dinner. I brush my teeth and change into my nightgown. I go to the bathroom one last time, and then my mother straps on the splints, placing a lighter blanket on top of me.

For three years, this is our nightly routine.

The splints minimize the time in the tub each morning, but still the pain continues. Most mornings it is a dull ache, but on days when it snows or rains I feel it in my bones. On days when it snows or rains, my mother reminds me to tell the gym teacher if it hurts to run or hop or do what it is he wants us to do.

"Remember," she says, helping me into my coat, "'juvenile rheumatoid arthritis.'"

But when the gym teacher asks us to pick partners or collect scooters from the back wall of the gymnasium, I don't want to sit on the sidelines. I want to play with the parachute, and the jump ropes, and the colorful square rugs. To avoid being left out, I learn to modify my movements: I sit cross-legged instead of kneeling, tie my shoe if he asks us to hop, move in sharp, zigzag patterns instead of running in dodge ball.

No one—not even the teacher—seems to ever notice.

Shortly after my sixth birthday, the pain goes away altogether. Each morning I wake and can stand.

"That's usually what happens with JRA," my doctor says, and I understand this as a victory, though I'm not sure what I've done.

On the way home from the hospital, my parents and I get ice cream at Freddy Hill Farms: tall waffle cones with lots of pastel scoops.

The arthritis is all I remember from my early childhood: the pain, and the baths, and the Lucky Charms.

By seven, the arthritis in my right knee went dormant, and so I identify those years with memories of other things—things not related to pain. I remember a birthday cake at my neighbor Michael's house when I was seven, a twenty-four pack of colorful markers at ten, a guinea pig that smelled like cedar. I remember junior high, a blue skirt my mother threw out because it was too short, a hula girl-shaped air freshener I bought for my first car. I remember high school graduation, and leaving Philadelphia, and moving two hours west to Gettysburg for college. I was eighteen then. I had forgotten all about the splints, the arthritis, the doctor with the foreign name.

But eight months later, on the sandy beach of Assateague Island in Maryland, I wake to a seizing pain in my knee.

It is the first weekend of summer. My college boyfriend and I have driven three hours south to celebrate by pitching a tent and cooking kielbasa over a fire. I am lying there at midnight, listening to thunder far off in the distance—thinking about the metal of the tent poles—when I feel it.

It has been twelve years since the swollen me I remember, and so I focus on my workouts, my runs, whether I pulled something or tore something else. I don't think about the arthritis at all.

"Wake up," I say at last to the breathing body beside me. I have been with him for six months, but he doesn't know about the splints, and the pain, and the tracks. "I hurt," I say.

He moans in his sleep, shifts against me in his sleeping bag. He doesn't say anything. "I mean it," I say, "it *hurts*. You run— is that what this is?"

The boy sighs. He rubs his tongue along the thick, dry line of his lips and says, "Eh."

It's too late for anything to be open, and my cell phone is tucked in the lower pocket of a book bag outside in the car, so I roll onto my back and try to keep my legs straight as I once did. The instinct is natural. I think about the morning, wait for the daylight, wait for offices to open. I focus on the sea, listen to the waves, breathe as the ocean breathes.

In the morning, I peel back the fabric of my sleeping bag to show the boy my right kneecap, swollen and thick in the hot white light. He carries me to the car, and I sit in the passenger seat as he dismantles our tent and rolls up our sleeping bags and carries our cooler to the trunk.

"Should we visit someplace local?" I ask, hoping we may be able to salvage what is left of our vacation. "Maybe we can come back once we've finished?"

"No," he says. He slams the trunk. "They might want to keep you. You might need your things."

On the three-hour drive back, I sit awkwardly, my feet propped on the dash. I try to find ways to apologize for what I've done. I joke that at least now it is summer, that in some ways my body couldn't have picked a better time to break. He leans over and squeezes my hand against his.

"It's okay," he says.

We stop at a gas station for a frozen fried chicken entrée to keep pressed against my knee.

In the emergency room that afternoon, a receptionist advises me to keep my leg elevated on a chair until the doctors can see me. "To ease the swelling," she says, though I can't imagine how any more fluid could possibly work its way into my joint. It is red, the skin pulled taut like a balloon filled with sand. It throbs.

While we wait, the boy reads to me from a copy of *Le Petit Prince* he keeps in the backseat of his car. I listen to the words on his tongue as they trickle off smoothly, his lips aligned.

"Ce qui embellit le désert," he says, "c'est qu'il cache un puits quelque part."

What makes the desert beautiful is that somewhere it hides a well.

When at last my name is called, the doctor takes an x-ray. He holds the image up to the light. "Nothing that I can help you with," he says.

He gives me a card for an orthopedist and a prescription for an anti-inflammatory. He has a nurse bring me a wheelchair to get to the car.

By the next morning, my kneecap is as thick as my thigh. I call my mother, and together we decide it is best I move home for the summer. She drives to Gettysburg and helps me pack my things. I say goodbye to my boyfriend, and then we drive the two hours east toward Philadelphia in silence. We sit in silence as the car moves.

At home, I visit one doctor and then another. I meet with a general practitioner, and then an orthopedist, and then a podiatrist. On several occasions, I mention my history with JRA.

"Not what this is," one doctor tells me. "That almost always disappears by the time a patient is eighteen. It's highly unlikely it would return."

For three weeks, my doctors and I consider runner's knee, patellofemoral pain syndrome, a stress fracture. The doctors draw blood to test for Lyme disease, autoimmune disorders, irregularities. I buy sneakers with arch support. I avoid stairs. The doctors continue to draw blood, feel my bones, remove enough fluid to fill a soda can.

"Twelve ounces, like a Mountain Dew," one says, holding it up to the light.

The doctors prescribe me anti-inflammatories and steroids. They make my kneecaps small but cause my cheeks to swell. The doctors give me crutches—to ease the impact of my weight on the joint—but it's not long before I give up. My arms aren't strong enough. My armpits ache from where the crutches push. Instead, I spend most of May on a mattress.

"It could be worse," my friends say when they visit. They sit on corners of my bed, their swimsuits soaking through their shorts from hours at the public pool. "You could," they laugh, "be a *total* cripple."

I do not tell them about the needles, the baths, the splints that set my legs ablaze.

In June, they stop visiting altogether. I spend the rest of summer watching reruns of *Big Love* on my bed, my leg propped up

AMY BUTCHER

with pillows, listening to the sounds of my brothers swimming in the pool below.

It is a rheumatologist who finally has my blood tested for a white blood cell count. The number is high. "Your body is fighting itself," he says. "You're a rare case—most children with JRA are free of the disease by eighteen. It never returns again. That's why it's called *juvenile.*"

He tells me that I am one of the select few who experience arthritic conditions after eighteen.

"At this point," he says, "it's considered full-blown rheumatoid arthritis." He clears his throat and tells me this is something I will have to manage for the rest of my life.

"Just like that?" I ask.

He looks at his clipboard. He rolls my pant legs to my thigh and squeezes the joint to feel for fluid. He removes another soda can's worth and writes a prescription.

"Come back once it's filled," he says. "We'll teach you how to use it."

"Isn't it digestible?" I ask. "Isn't it just a pill?"

"Enbrel," he says. I think of my doctor and his foreign words. "It's a shot," he says. "It gets injected once every week in your stomach, or your thigh."

Back in the waiting room, I pay the co-pay through tears, remembering the splints, the needles, the yellow board, the baths. All of history rushing back.

The first time I inject myself, I'm in the rheumatology office with a nurse who inspects my every move. "That's right," she says as my finger pushes down on the trigger of the needle. "Good."

When I'm done, I hold a cotton ball to the injection site. She peels back the label of a Band-Aid, affixing it over the small nub of cotton.

"Not so bad?" she says, but it is. The medicine stings, and the injection site grows a red ring the size of a grapefruit. And then there are the risks: that Enbrel increases my likelihood of developing lymphoma, that I have to go off it when I become pregnant, that I need to get a flu shot every year because if I don't, the infection could get out of hand.

"That's how people die," my doctor says. "The medicine suppresses your immune system so your body doesn't fight your body. Without a vaccine, the flu could be fatal."

I nod like I understand because I want my legs to work again. But at night on the phone, I tell the boy about the risks.

"The flu," I say. "Do you believe that? The *flu.*"

He speaks to me with soothing words, but there is an uneasiness in his voice. After we hang up, I lie awake in the darkness and wonder if he wants to fall in love with someone who could get so sick, who could become immobile while pregnant, who could die of something so simple as the flu. He is a boy who runs, a boy who lifts weights and travels to countries I can't even point to on a map.

I lie quietly and wonder if I will slow him down—which, in time, I do.

Four years later and over a thousand miles away, I invite my friends over to my apartment. It is the eve of my twenty-third birthday. We spend the night drinking wine and eating expensive cheeses from the neighborhood co-op. There is camembert and smoked gouda, Drunken Goat and lavender-crusted brie. There are paper lanterns. There is a playlist of Edith Piaf and Frank Sinatra.

"This is an adult birthday party," I say, feeling young even as I say it.

Halfway through the night, a friend taps me on the shoulder. She informs me that I am out of toilet paper. She is drunk, her lips stained red.

"We've got a problem," she says.

"No, there's more," I say. I uncork a bottle of Cotes du Rhône. "Check the cabinet under the sink."

Minutes later, my friend reemerges. She giggles, bumping against the tiered fruit stand. "Okay?" I ask her.

"I found some," she says. "But what's with that red canister marked with a biohazard sign?" She laughs, pours herself another glass, steadies the wobbling apples beside her. "What don't I know about you?" she says. "Why are you dealing with biohazardous material?"

The biohazardous material is the collection of used needles. Each week, I inject the medicine, feel the sting, and then discard the needle in the red plastic bin. When it fills, I must make an appointment and drive twenty-two miles outside of the city. I sit in a dusty field for half an hour behind a line of cars until it is my turn, and a man wearing latex gloves opens my car door for me and takes the bin for proper disposal.

I shake my head. I reach for the oranges. "There's a can of cranberry cocktail in the cabinet," I say. "Will you grab it for me? Punch bowl's dry."

What my friend doesn't know is what so many people don't know, because there is no need to tell them. The needles make the disease invisible. They make the joints look normal. The pain is still there from the months of swelling—the cartilage worn away the way bicycle chains lose their grease—and I feel it when I climb steps, lean, stand for periods of time. I feel it when I stand up after sitting for more than ten minutes. I feel it when it rains. But as long as the needle goes in my thigh, I look like every other twenty-three-year-old. The secret can be mine, and I prefer it to be mine alone. I walk and bike and swim. I pretend to have young skin, young bones, young joints, young everything. The needles make it easy to forget and easier still for others not to know.

But they don't fix everything—don't give me back the memories, don't re-create what I do remember. There will always be the tub, and the splints as straight as rails, and the fish who move fluidly beneath thin plastic while a woman reconstructs my childhood. There are moments of stiffness on buses or trains. There are moments when assimilation is impossible.

And when the yoga teacher presses up on my abdomen, says, "Push *deeper*, push *farther*," and I feel her warm breath on my ear, the grating in my knee, I remember all of this.

ANGELA PELSTER

Once, Then

ONCE, I WASHED his feet. He washed mine. Because once, they told us, Christ took off his coat and tied a towel around his waist and knelt, cradled the disciples' cracked heels in his hands like baby birds, and cleaned the rough and ticklish places together. "Go and do likewise," we were told. So I took his feet in my hands and washed them clean. Wanting grace. Or God. Or miracle. Or show. All of it. And he did the same with mine. Those nights we slept sole to sole, grounded on the other.

Once, one night, in a backyard pool, we swam together, and everything touched, every limb found heat, ecstasy shushed quiet by the other in the dark. And my foot was sliced deep with his nail. I bled into the cold water, hurt and laughing, showed him later, for years, the scar over the smooth blue veins that he had given me.

Once, those first days, we held her new. Washed her in a plastic tub on the kitchen table. Cleaned the dried blood from her ears, the folds of skin along her neck, the space between each finger. She wailed when I took her toes in my teeth, nibbled the nails to keep her safe from scratches. Pale limbs flailed, heart, blind to any kindness but my depths. But, oh, her rounded soles so mint condition. That foot, I awed, that one that had pushed inside me nights while I grew her.

Then, those nights, time circled while I waited. Until late black, nearly morning, he crept back in. Sun almost rising. The house not lit.

I listened as he went down the stairs, into the basement. I do not know why. He took to pissing in the floor drain. All that vodka down the drain. The secret bottle stashed in the rafters. Feet soaked, soaked legs as he topped the stairs, a wet trail below. And me at the kitchen table waiting.

A broken tree about to topple. Limbs bent, eyes akimbo.

The rustle of baby in her crib down the hall.

The click of a light flicked on, as we blinked under it.

This, some inquisition; this, some role I did not know how to play.

Soon, I would sit beside him on the couch. He would confess to the women like groceries, a list. Here and here, with her and her. I listened, traced my finger along the smooth skin of his arm, one smooth place, so smooth, so unable to wake to it.

I broke my toe. His broke too. When I kicked at him, he said. He said I kicked him and broke his toe. When my hands flew at the things he had done and he picked me up and slammed me into the new white fridge to stop my wild arms and left his purple fingerprints on me. And toes broke, snap, snap, like the other things.

But first, there was this: this first, this night, when he stood in the kitchen and swayed, when my hands moved to their theology: to pull a chair from the table, fill a bucket with water, and kneel to wash.

His face sweated shame, mine twitched its grasping. Bodies, only flesh, my feet, already racing, hurtling toward the dark, and the done, and the breaking. There was no stopping it, though I would not know it then, or the why of it, any of it, ever, yet, still. Twelve years into after. Nothing to say of it but this: Dear humans, both, on the edge of your making, chin up. It will not cease to hurt; it will one day cease to matter.

Contributors

AMY BUTCHER is an essayist and author of *Visiting Hours*, a memoir that earned starred reviews and praise from the *New York Times Sunday Review of Books*, NPR, the *Star Tribune*, and *Kirkus Reviews*. Her essays have appeared recently in the *New York Times*'s "Modern Love" and the *Washington Post*.

WENDY CALL co-edited *Telling True Stories: A Nonfiction Writers' Guide* and wrote *No Word for Welcome*, winner of Grub Street's National Book Prize. Artist residencies she completed at Harborview Medical Center and the American Antiquarian Society inspired her essay "Beautiful Flesh." She teaches at Pacific Lutheran University and lives in Seattle.

STEVEN CHURCH is the author of five books of nonfiction, most recently *One With the Tiger: Sublime and Violent Encounters Between Humans and Animals*. He is a founding editor and the nonfiction editor of the *Normal School* and of a forthcoming anthology of essays due out from Outpost19 in the spring of 2018. He is the coordinator of the MFA program at Fresno State.

SARAH ROSE ETTER is the author of *Tongue Party* (Caketrain Press). Her work has appeared or is forthcoming in the *Collagist*, *Black Warrior Review, Salt Hill Journal,* and more. She is a co-founder of the TireFire Reading Series and a contributing editor at the *Fanzine*. www.sarahroseetter.com.

MATTHEW FERRENCE lives and writes at the confluence of Appalachia and the Rust Belt. He is the author of a book of cultural criticism, *All-American Redneck,* and is working on a project that explores the geographical and conceptual limits of Northern Appalachia. He teaches at Allegheny College.

HESTER KAPLAN is the author of novels and short story collections, including *The Edge of Marriage,* winner of the Flannery O'Connor Prize for Short Fiction. Her stories and nonfiction have been widely published and anthologized, including in *The Best American Short Stories* series. She teaches in Lesley University's MFA Program in Creative Writing.

SARAH K. LENZ's nonfiction has appeared in *New Letters, Crazyhorse, Colorado Review,* the *Fourth River,* and elsewhere. "The Belly of Desire" is from her collection-in-progress, "Lightning Flowers." In 2015 she received the New Letters Readers' award in nonfiction. She holds an MFA from Georgia College.

LUPE LINARES is an assistant professor at Ball State University, where she teaches courses in American literature and composition. She no longer carries her dad's old license around because it was stolen, along with her purse, at the Coffee House in Lincoln, Nebraska. The world is a cruel place.

JODY MACE is a freelance writer living in North Carolina. Her essays have appeared in *O Magazine; Brain, Child;* the *Washington Post;* and many other publications, as well as several anthologies. She is a regular contributor to FullGrownPeople. com, where this essay originally appeared.

DINTY W. MOORE is author of *The Story Cure: A Book Doctor's Pain-Free Guide to Finishing Your Novel or Memoir,* the memoir *Between Panic & Desire,* and other books. He has published essays and stories in the *Southern Review,* the *Georgia*

Review, Harper's, the *New York Times Sunday Magazine,* and elsewhere.

ANGELA PELSTER's essay collection *Limber* won the Great Lakes Colleges Association New Writer Award and was a finalist for the PEN/Diamonstein-Spielvogel award. Her work has appeared in the *Kenyon Review, Hotel Amerika, Granta, River Teeth, Seneca Review,* and the *Gettysburg Review* among others. She teaches creative writing at Hamline University.

MATT ROBERTS is a founding editor of the *Normal School* literary magazine. His work has appeared in *Isotope, Post Road, Ecotone, Ninth Letter,* on National Public Radio's *Morning Edition,* and in the anthology *No Near Exit: Writers Select Their Favorite Work from Post Road Magazine* (Dzanc Books).

PEGGY SHINNER is the author of *You Feel So Mortal,* a collection of essays on the body (University of Chicago Press), which was long-listed for the 2015 PEN/Diamonstein-Spielvogel Award for the Art of the Essay. Currently, she is at work on a book about shame.

SAMANTHA SIMPSON currently teaches English at the Cambridge School of Weston and leads creative writing workshops at the *Kenyon Review* Young Writers Program. Her best story and essay ideas arrive when she's five miles into a long run.

FLOYD SKLOOT's work has won three Pushcart Prizes, the PEN USA Literary Award, and been included in *Best American Essays, Best American Science Writing, Best Food Writing,* and *Best Spiritual Writing.* In 2010 *Poets & Writers* named him "One of 50 of the Most Inspiring Authors in the World."

DANIELLE R. SPENCER is a narrative medicine faculty member at Columbia University and previously worked as artist/musician David Byrne's art director. She is co-author of *The*

Principles and Practice of Narrative Medicine and her work has appeared in the *Lancet, Creative Nonfiction, Esopus,* the *Hungarian Review*, *Wired*, and *The Routledge Companion to the Philosophy of Medicine.*

KATHERINE E. STANDEFER's writing appears in *The Best American Essays 2016,* won the 2015 Iowa Review Award in Nonfiction, and has been nominated for a Pushcart Prize. In Tucson, she teaches intimate, independent writing classes on sexuality, illness, and trauma. Her current book project, "Mountains in My Body," traces the global supply chain of her defibrillator.

KAITLYN TEER holds an MFA from Western Washington University, where she served as managing editor of *Bellingham Review*. Her essays have received awards from *Fourth Genre* and *Prairie Schooner*. Other work appears in *Camas, Midwestern Gothic, Sweet,* and *Entropy.*

SARAH VIREN is a writer, translator, editor, and former newspaper journalist. Her work has appeared in *Agni*, the *Iowa Review, Gettysburg Review, TriQuarterly,* and other magazines. Her essay collection *Mine* won the River Teeth Literary Nonfiction Book Prize and is forthcoming from the University of New Mexico Press. She is an assistant professor at Arizona State University.

VICKI WEIQI YANG currently lives in Washington, D.C., and will begin a Ph.D program in sociology in fall 2017. At the time of writing this essay, she had no technical experience with ethnographic field notes.

Permissions

"Taking Shape," by Amy Butcher. First published in *Michigan Quarterly Review*, volume 1, issue 2, Spring 2011. Copyright © 2011 by Amy Butcher. Reprinted by permission of the author.

"Beautiful Flesh," by Wendy Call. First published in the *Georgia Review*, volume 69, number 3. Copyright © 2015 by Wendy Call. Reprinted by permission of the author.

"Speaking of Ears and Savagery," by Steven Church. First published in *Creative Nonfiction*, 45. Copyright © 2012 by Steven Church. Reprinted by permission of the author.

"The Spine," by Sarah Rose Etter. First published in *Fanzine*, 25.03.14. Copyright © 2014 by Sarah Rose Etter. Reprinted by permission of the author.

"Mos Teutonicus," by Matthew Ferrence. First published in *Colorado Review*, volume 41, number 3. Copyright © 2014 by Matthew Ferrence. Reprinted by permission of the author.

"The Private Life of Skin," by Hester Kaplan. First published in *Southwest Review*, volume 91, number 2. Copyright © 2012 by Hester Kaplan. Reprinted by permission of the author.

"The Belly of Desire," by Sarah K. Lenz. First published in *Crazyhorse*, 89. Copyright © 2016 by Sarah K. Lenz. Reprinted by permission of the author.

Acknowledgments

MY DEEPEST THANKS go first to the writers who agreed to be part of this project, and next to the editors who helped me find them—especially Hattie Fletcher, Sophie Beck, and Robert Atwan, true compatriots all the way.

I'm extraordinarily grateful, as well, to the entire staff at the Center for Literary Publishing, who kept all our other plates spinning while I focused on this one. I'm particularly thankful for Cedar Brant's hand in all of this.

I also owe many thanks to the English Department at Colorado State University, my home for these many years, but especially to my colleagues in Creative Writing and Creative Nonfiction.

And finally, thank you to Steven Schwartz, Emily Hammond, and Arne G'Schwind for their wisdom, support, and encouragement, so generously given when I needed it most.

Typesetting by Cedar Brant
at the Center for Literary Publishing
at Colorado State University.
Proofreading by Dana Chellman,
Cory Cotten-Potter, Michelle LaCrosse,
& Morgan Riedl.
Cover art by Travis Bedel.
Cover design by Stephanie G'Schwind.